# MATERIALS IN CONSTRUCTION

**G D Taylor**
BSc Hons, PhD, CertEd, MCIOB

# Materials in Construction
**Second Edition**

Longman
Scientific &
Technical

**Longman Scientific & Technical**
Longman Group UK Limited
Longman House, Burnt Mill, Harlow
Essex, CM20 2JE, England
*and Associated Companies throughout the world*

First published 1985
Second edition 1994

**British Library Cataloguing in Publication Data**
A catalogue entry for this title is available from the British Library

ISBN 0-582-21431-9

Set by 8 in 10/11pt Times
Produced by Longman Singapore Publishers (Pte) Ltd
Printed in Singapore

# Contents

## Chapter 3: Bricks, building blocks and mortars 86

## Chapter 4: Dampness in buildings 112

# Acknowledgements

We are grateful to the following for permission to reproduce copyright material:

The Controller of Her Majesty's Stationery Office for figures 2.11–2.15 and tables 2.8 & 2.9 from *Design of Normal Concrete Mixes* BRE and figure 6.12 from *The Strength Properties of Timber* BRE Princes Risborough Laboratory, all Crown Copyright; British Standards Institution for figures 6.8, 6.10, 6.19 & 6.20 and tables 2.2, 2.11–2.14 & 7.1–7.5, all *BS Specifications*, complete copies may be obtained from the BSI, Linford Wood, Milton Keynes, MK14 6LE.

# Chapter 1

# Materials in construction

## Introduction

Modern buildings often comprise vast numbers of components, many pre-manufactured and simply assembled on site, while some, such as concrete, are manufactured *in situ*. In each case, satisfactory operation of the building as a whole depends on the performance of the materials from which its components are made as well as on how they interact with each other in the building.

Before assessing the suitability of any one material for a given situation, the 'performance' requirements for that situation must be identified. Such requirements might include:

- *Structural safety* – the ability to withstand stresses resulting from gravity, wind, thermal or moisture movement, or other sources.
- *Health/safety* – there should be no risk to health due to chemical or physical effects of the material both during and after construction.
- *Fire* – the material must behave acceptably in resisting fire spread, release of dangerous substances in fire and retaining satisfactory structural stability.
- *Durability* – the material should fulfil the above performance criteria as required for the planned lifetime of the building.

There are a number of other important performance requirements: for example, comfort, resistance to weathering, serviceability and appearance which will be referred to as appropriate when individual materials are described.

An important feature of most buildings is that they are often unique - for example in situation, composition or function. Hence, they are each subject to a different combination of the parameters listed above, For each building, the performance requirements must be identified and then acceptable methods of achieving these requirements sought. The subject is, therefore, complex and the construction team will benefit from a thorough knowledge of materials properties, applications and limitations.

In addition to the above performance requirements of materials within the finished building, the following might also have to be considered:

- Availability/cost.
- Ease with which material can be incorporated into the building (buildability).
- Environmental aspects – for example, energy demand of the material during manufacture and ability to conserve energy in use.

## Specifications and standards

Most new buildings are produced with the aid of some written specification, which may often take the form of detailed annotations to drawings, a bill of quantities, or separate documents which are part of the contract. Some form of written specification is highly desirable because it communicates to the construction team exactly what is required and provides for the clients a basis on which unsatisfactory performance may be identified and remedied.

Specifications are almost invariably centred around accepted standards which provide a simple, convenient means of specifying performance levels. Most standards represent basic performance levels; when higher levels are required these may have to be drafted carefully for a given specification. Standards measure performance in a carefully defined reproducible manner and they are subject to change as understanding of materials properties increases, experimental techniques improve and performance requirements evolve.

## British Standards

These began to be produced from the beginning of the twentieth century and now cover a vast range of materials and components, including most of those used in the construction industry. They are produced by committees representing manufacturers, researchers, users and government organisations. It is important to appreciate that manufacturers can claim compliance with a British Standard without independent tests being carried out. The best evidence of conformity is the Kitemark (Fig. 1.1) or other evidence of independent assessment. With the advent of Eurostandards,

Fig. 1.1 The Kitemark as evidence of BS conformity by an independent test authority.

British Standards now operate only on a care and maintenance basis. They will eventually be replaced by their European counterparts which have validity in Europe as a whole.

## Agrément certificates

These are produced by the British Board of Agrément and generally relate to new products or techniques not covered by a British Standard. Being widely accepted by local authorities, they represent a means of obtaining relatively rapid approval and hence are an encouragement to innovation, though certificates are valid for a limited time period prior to preparation of British or other standards. A European Union of Agrément (UEAtc) operates on a similar basis in other European countries.

## European codes and standards

These provide a mechanism by which construction products and materials can be traded and used freely throughout the European Common Market. They are produced by Technical Committees of the European Committee for Standardisation (CEN), which are of international composition reflecting the viewpoints of member states. Although December 1992 was considered the implementation date for these standards, their introduction and validity will be a gradual operation since, at present, many standards are incomplete. As with British Standards, there is a need for recognised testing and validating authorities in each member state and this infrastructure will inevitably take time to introduce.

So far there are nine Eurocodes:

1. General and actions
2. Concrete
3. Steel

4. Composite (concrete/steel)
5. Timber
6. Masonry
7. Foundations
8. Seismic design
9. Aluminium

Each Eurocode can be regarded as a code of practice, being backed up by European Standards (ENs). These standards are based on six 'essential requirements' relating to health, safety and energy conservation. They are:

- stability
- safety in fire
- safety in use
- health
- noise
- heat retention

As with British Standards, European Standards must be developed with great care, representing the viewpoints of various 'interested' parties. They are therefore produced first as pre-standards or 'ENVs' which have no legal status and are then followed by the full standard after a period of about three years. Any product which complies with the essential requirements given in the standard is eligible to carry a 'CE' mark (Fig. 1.2).

Any standard will require a level of 'attestation' which demonstrates that it has been adequately assessed. For critical aspects of behaviour, third party attestation will be necessary. In less critical aspects the manufacturer may be able to monitor the product, although quality systems such as BS 5750 will need to be operated. The importance of European standards can be judged from the Building Regulations (1991), Regulation 7 stating that:

> An EC marked material can only be rejected by the Building Control Authority or Approved Inspector on the basis that its performance is not in accordance with its technical specification. The onus of proof in such cases is on the Building Control Authority or Approved Inspector, who must notify the Trading Standards Officer.

Fig. 1.2 The 'CE' mark, indicating compliance of a product or material with its appropriate Eurostandard.

## Quality of products (BS 5750)

*Quality* can be simply defined as 'fitness for purpose', although in practice it is necessary to interpret the term precisely in each situation and then to decide how 'quality' requirements should be implemented, bearing in mind that there will always be a cost implication as target quality levels rise. Hence, the first stage in considering a specification is to arrive at an appropriate/realistic target for quality. Some factors to be considered when arriving at such a target level for a specific item are:

1. What are the possible failure modes?
2. What are the consequences of failure in safety terms?
3. How easy is it to inspect/maintain the item?
4. How easy/costly would it be to replace the item if it failed?

### *Quality control*

This is the practical procedure which assists in the production of a quality product. In the contexts both of items produced in a factory or on site, there are numerous opportunities for defects to arise whether by machine or human error. Quality control attempts to identify these possible events and then implements procedures to avoid them if possible and to detect and rectify them should they occur. The mere fact that a process is being monitored is likely to assist in reducing the number of human errors, especially if penalties are imposed for mistakes. It is important to appreciate that in this technological age, personnel rarely have a full understanding of the manufacturing process in which they are involved, or of all the consequences of any errors. In this situation the importance of adequate supervision will be evident. In any one situation the level of supervision invoked must be decided by consideration of the performance and reliability required of the item being produced, together with the overall cost of a failure, including damage to reputation, compared with the cost of providing a given level of quality control.

### *Quality assurance*

This involves the operation of a comprehensive system of quality control, including employment of a quality manager to oversee the maintenance of quality standards and to keep systematic records of every part of a design, production, or other process.

Although companies are not required to register quality assurance schemes with a third party such as the BSI or other approved certification body, there are obvious advantages, in public recognition terms, of doing this. The certification bodies themselves must be accredited and this is organised through the National Accreditation Council for Certification Bodies (NACCB). The BSI has the most quality assurance registrations at the present, though other important QA Accreditation bodies in the constuction area include Lloyd's Register Quality Assurance, the Quality Scheme for Ready Mixed Concrete and the UK Certifying Authority for Reinforcement Steels. The BSI 'Kitemark' can currently only be used in conjuction with products which are manufactured to a BS 5750 quality assurance scheme. Quality assurance is more readily applied in factory situations where operations are more mechanised and repetitive than in site situations, though designers, building and other contractors, and those providing services are rapidly moving towards quality systems of this type, having regard to the benefits – both in lower failure rates and marketing advantages – of quality assurance.

## Materials performance and its measurement

The main headings under which performance might be assessed are given earlier in this chapter. There follows an amplification of these areas in order that the critical parameters in any one material might be identified and controlled.

### Structural safety

There are many aspects of structural performance, the main ones being as listed below.

*Strength*
This may be defined as the ability to resist failure or excessive deformation under stress. There are several types of stress (Fig. 1.3) and in any one situation the stress pattern should be identified and performance checked. Compresssion testing needs special care because:

- Slender specimens may tend to buckle (bend).
- Friction with the steel plate of the testing machine affects the results.
- Compression test results depend on the **shape** of the samples, since they may fail by induced **shear** stresses.

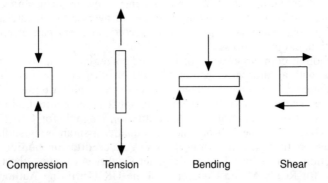

Compression     Tension     Bending     Shear

Fig. 1.3 Basic load configurations.

Note that bending produces compression, tension and shear stresses within the material. All stresses should be measured in newtons (N) per square millimetre (mm²). Note that fracture is not necessary for a strength 'failure'; for example, steel for practical purposes, has failed when it 'yields', though yielding does not initially damage the metal. There are, on the other hand, many materials – ceramics, for example – where fracture is the first indication that strength has been exceeded.

## Stiffness
This term normally relates to **elastic** deformation, that is, deformation which is recovered when the load is removed. High deformations, even if elastic, may cause problems, for example, unsightly appearance or failure of plaster coatings. In practice, therefore, they should be checked. The ability of a material to resist plastic deformation is referred to as 'stiffness', normally measured by:

$$\text{Elastic modulus} = \frac{\text{Applied stress}}{\text{Strain caused by that stress}}$$

Since strain has no units (= fractional change of length), $E$ values have the units of stress, usually $kN/mm^2$ ($1 \ kN/mm^2 = 1000 \ N/mm^2$).

## Toughness
This is the ability to absorb energy by impact or sudden blow. Strong materials are not always tough – for example, cast iron; while relatively weak materials can have high toughness – for example, leather.

## Hardness
This is resistance to indentation and is relevant to floor and wall surfaces (Fig. 1.4). Hardness depends on a combination of strength and stiffness properties.

## Creep
This is the effect of long-term stress under which some materials gradually deform and eventually break. Materials subject to creep are timber, clay, lead, concrete, thermoplastics and, to a small extent, glass.

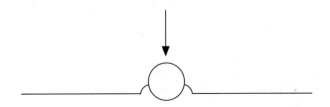

Fig. 1.4   Hardness measurement by an indentation test.

8

*Fatigue*
This is the effect of load reversals such as vibration which lead to failure at relatively low stresses. All materials are subject to fatigue effects and in some situations – for example, roads or floors subject to heavy moving loads, or machine frames – fatigue may be the critical factor in design.

## Health and safety

Public awareness of health and safety issues is increasing rapidly and the construction industry has come under scrutiny both from the point of view of safety of the construction team and the general public during construction process and because the finished building might present a hazard to the occupants. A detailed examination of all the aspects of health and safety relating to building materials is beyond the scope of this book, though Table 1.1 indicates some of the possible safety hazards posed by materials, together with recommendations for overcoming them.

## Fire

Combustion is in essence a simple process involving chemical reaction of a fuel (combustible material which is usually organic, contains carbon) with oxygen. To initiate the process, heat or a source of ignition is essential, though once started, many combustion processes are self-sustaining because heat is a by-product. The situation is illustrated by the triangle of Fig. 1.5.

To prevent or extinguish a fire, it is only necessary to remove one of the three essential ingredients. For example, fire risk is greatly reduced as the quantity of combustible material in an enclosure (fire load – Building Regulations) is reduced. Some extinguishers work by surrounding the fire with a gas such as a 'halon', containing the halogens chlorine, fluorine or bromine which interfere with the combustion process or prevent oxygen

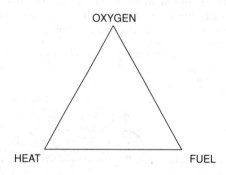

Fig. 1.5   The fire triangle. If any one ingredient is missing the fire cannot start.

Table 1.1   Risks associated with toxic materials used in the construction industry

| Substance | Situation | Risk | Remedy |
|---|---|---|---|
| Lead | Formerly in paints, pipes solders | Health risk if ingested, especially to children | Remove existing lead pipework. Specify lead – free paints and solders |
| Radon gas | Diffuses from some granite rocks through ground floor into houses | Radioactive gas may lead to risk of lung cancer | Isolate ground floor by membrane or prevent ingress by positive internal pressure |
| Formaldehyde gas | Present in some foams, e.g. cavity fills; glues as in chipboard | Can cause nausea | Ventilate new dwellings |
| Chlorofluoro-carbons (CFCs) | Used in refrigerants, air conditioning systems, propellants for aerosols and foaming agents for some plastics | Depletes ozone layer, hence contributes to global warming and increases UV radiation at earth's surface | Check specifications for products which might contain CFCs. Less harmful substitutes now available |
| Wood preservatives | During storage, transport and at or soon after application | Risk to human, animal or plant life | Use only in safe situations under strict control |
| Asbestos | Not currently used but may be present as insulation or other forms in buildings | Risk of lung cancer | Arrange for safe removal if found |

access. Water extinguishes fire by removal of heat, greatly assisted by vaporisation of water which is a highly endothermic reaction, consuming very large quantities of heat.

The calorific value of a fuel measures the heat output on total combustion and does vary greatly within the range of organic materials. For example, wood, coal, paper, petrol are all in the range 15–50 MJ/kg. The risk in practice varies greatly, however, being greatest with liquid or gaseous fuels which mix very easily with oxygen and least with bulk solid materials such as large timber sections where oxygen access is limited.

Hence fire is a highly complex subject and this is reflected in practice by the numerous BS tests for the different aspects of fire. The most important of these are illustrated in Figs 1.6 to 1.11.

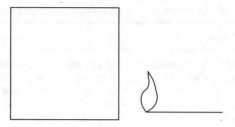

Fig. 1.6   Non-combustibility: BS 476, Part 4; ignitability: BS476, Parts 12/13; heat emission: BS 476, Part 11.

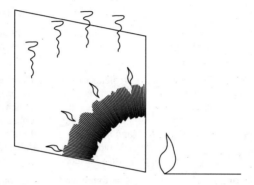

Fig. 1.7   Surface spread of flame: BS 476, Part 7.

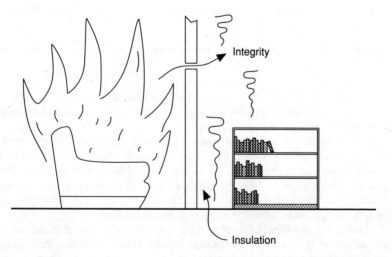

Fig. 1.8   Fire resistance, integrity and insulation (door/partition): BS 476, Parts 20 to 24.

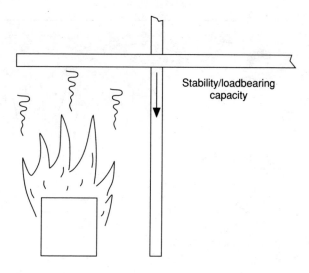

Fig. 1.9    Fire resistance, stability (wall): BS 476, Parts 20 to 24.

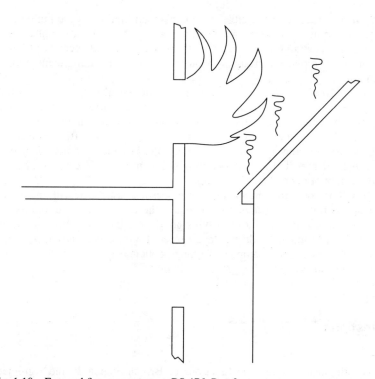

Fig. 1.10    External fire exposure test: BS 476, Part 3.

Fig. 1.11   Effects of smoke: BS 476, Part 31.

## Durability

A material may be said to be durable in any one situation if it fulfils all its performance requirements, either for the planned lifetime of the building, or for a shorter defined period where this is acceptable – for example, where replacement is straightforward, where there is a severe cost penalty of longer life or where replacement arising from changing user requirements may be desirable.

It is often very difficult to predict the durability of individual components. Also, since failures in a very small proportion of the items in use may be unacceptable, the only safe course of action may be to 'over-design' them so that materials in the worst likely situation should be satisfactory. In consequence, many buildings last much longer than their design lifetime. Table 1.2 summarises some of the main modes of deterioration of major materials groups.

It will be noted that the majority of modes are associated with the effects of water. A further important point is that if these modes of deterioration are avoided, as should be the case with good practice, long life should be achieved; most materials do not 'age' except by these specific mechanisms. There are, for example, samples of timber, concrete and metals that are over 1000 years old and are still in good condition.

## Questions

1.  Distinguish between the functions of British Standards and Agrément certificates.

Table 1.2  Modes of deterioration of the major materials groups

| Type | Form of deterioration | Recommendation |
| --- | --- | --- |
| Metals (non porous) | Surface deterioration by electrolytic corrosion in damp conditions. Severity depends on metal type and situation | Use resistant metal such as stainless steel or apply coating to exclude moisture |
| Bricks, stone concrete (porous) | Moisture penetration followed by frost or chemical action. Deterioration at or below surface | Use non-permeable forms; weatherings to shed water, impermeable coatings to preventwater/chemical penetration |
| Timber and timber products | Fungal attack (damp conditions) insect attack (normal conditions) | Exclude water. Apply preservatives |
| Polymeric materials | Ultraviolet degradation | Protect from sun, e.g. carbon black, or use indoor/underground. Use stabilisers or surface coatings |

2. Explain the difference between the terms *strength* and *stiffness*. Give examples of materials of
   (a) high strength and high stiffness
   (b) low strength and high stiffness
   (c) high strength and low stiffness
   (d) low strength and low stiffness.
3. A material is to be used to clad the entrance hall to a large public building. Indicate three aspects of its fire performance which should be considered.
4. Explain the difference between the terms *quality control* and *quality assurance*.

## References

### British Standards

BS 476: *Fire test on building materials and structures.*
    Part 3: 1975, *External fire exposure test.*

Part 4: 1970 (1984), *Non-combustibility test for materials.*

Part 7: 1987, *Method of classification of the surface spread of flame for products.*

Part 11: 1982 (1988), *Method for assessing the heat emission from building materials.*

Parts 20 to 22, *Determination of fire resistance.*

Part 31, *Methods for measuring smoke penetration through doorsets and shutter assemblies.*

Part 31.1: 1983, *Method of measurement under ambient temperature conditions.*

BS 5750, *Quality systems.*

# Chapter 2

# Concrete

Concrete is essentially a mixture of cement, aggregates and water. Other materials added at the mixer are referred to as 'admixtures'. Materials based on cement have the following general attractions:

- Low cost.
- Flexibility of application – for example, mortars, concretes, grouts, etc.
- Variety of finishes obtainable.
- Good compressive strength.
- Protection of embedded steel.

The following are some disadvantages/problem areas:

- Low tensile strength/brittleness.
- Rather high density (though lower density types are available).
- Susceptibility to frost/chemical deterioration (depending on type).

## Cements

The most important are Portland cements; so named because of a similarity in appearance of concrete made with these cements to Portland stone. Portland cements are hydraulic – that is, they set and harden by the action of water only. They do not rely on atmospheric action or drying.

Portland cements are made by heating a finely divided mixture of clay or shale and chalk or limestone in a kiln to a high temperature – around 1500 °C – such that chemical combination occurs between them. About 5 per cent gypsum (calcium sulphate) is added to the resulting clinker in order to prevent 'flash' setting and the final stage involves grinding to a fine powder. ⱱ

Although the familiar grey powder may appear to have a high degree of uniformity, it is important to appreciate that Portland cement is a complex combination of the minerals contained originally in the clay or shale and the calcium carbonate which constitutes limestone or chalk. The chief compounds are listed below, together with commonly used abbreviations.

Clay or shale

| | | |
|---|---|---|
| $SiO_2$ | Silica (silicon oxide) | abbreviated S |
| $Fe_2O_3$ | Ferrite (iron oxide) | abbreviated F |
| $Al_2O_3$ | Alumina (aluminium oxide) | abbreviated A |

Limestone or chalk

$CaCO_3$ (calcium carbonate) on heating gives
$\quad\quad\quad$ CaO (quicklime) $\quad\quad\quad\quad\quad\quad\quad$ abbreviated C

Chemical analysis or microscopic examination of cement clinker shows that there are four chief compounds present. Table 2.1 summarises the properties of these compounds, including their heat emission on hydration and typical percentages in the most commonly used cement, known as 'ordinary Portland cement' and often abbreviated OPC. The percentages do not add up to 100 because small quantities of other compounds are also present.

The properties of Portland cements vary markedly with the proportions of the four compounds, reflecting substantial differences between their individual behaviour. The proportion of clay to chalk (usually approximately 1:4 by weight respectively) must be very carefully controlled during manufacture of the cement, since quite small variations in the ratio produce relatively large variations in the ratio of dicalcium silicate to tricalcium silicate. A greater proportion of chalk favours the formation of more of the latter, since it is richer in lime. This would lead to a cement with more rapid early strength development. The tricalcium aluminate and the tetracalcium aluminoferrite contribute little to the long-term strength or durability of Portland cements but would be difficult to remove and, in any case, are useful in the manufacturing process since they act as fluxes, assisting in the formation of the silicate compounds.

There will also be variations in cement properties from one works to another as a result of differences in the composition of the clay or shale so that, if maximum uniformity is to be achieved in a concreting job, cement from the same works is essential.

The properties of a cement are dependent not only on its composition but also on its fineness. This is because cement grains have very low solubility in water, so that the rate of reaction and hence setting and hardening increases as the surface area of grains increases – that is, as they become finer. Fineness is measured by the term 'specific surface', which is defined as:

Table 2.1 Properties of the four chief compounds in Portland cements

| Name | Abbreviation | Approximate percentage in OPC | Properties | Heat of hydration (J/g) |
|---|---|---|---|---|
| Dicalcium silicate | $C_2S$ | 20 | Slow strength gain – responsible for long-term strength | 260 |
| Tricalcium silicate | $C_3S$ | 55 | Rapid strength gain – responsible for early strength (e.g. 7 days) | 500 |
| Tricalcium aluminate | $C_3A$ | 12 | Quick setting (controlled by gypsum); susceptible to sulphate attack | 865 |
| Tetracalcium aluminoferrite | $C_4AF$ | 8 | Little contribution to setting or strength; responsible for grey colour of OPC | 420 |

$$\frac{\text{Surface area of the grains in a sample}}{\text{Mass of that sample}}$$

The units of specific surface are $m^2/kg$, that of ordinary Portland cement (OPC) being required by earlier editions of BS 12 to be not less than $225\,m^2/kg$, though in practice it is usually in the range $350–380\,m^2/kg$. The current edition of BS 12 (1991) does not specify a minimum fineness for OPC; indeed the standard, which has been produced with the developments in European standards in mind, does not use the term 'ordinary Portland cement' though the term is still used by cement manufacturers to describe Portland cements of strength class 42.5 N as described in the standard. This strength is based on mortar cube tests .

Ordinary Portland cement is the cement best suited to general concreting purposes. It is the lowest-priced cement and combines a reasonable rate of hardening with moderate heat output. Other types of cement are, however, available, each being recommended for specific applications. The chief variants on ordinary Portland are as follows:

## Rapid-hardening Portland cement

This could be obtained by increasing the $C_3S$ content as explained above but is normally obtained from OPC clinker by finer grinding (about $450\,m^2/kg$) and is covered by the same standard (BS 12) as OPC. Again whereas earlier editions of BS 12 required a minimum fineness of $325\,m^2/kg$, the current edition does not specify a value. Rapid-hardening Portland cement (RHPC) tends to set and harden at a faster rate than OPC. The rate of set is controlled by the addition of slightly more gypsum during manufacture, since it is the hardening properties and heat emission rather than setting rate which form the basis of applications. Early strength development is considerably higher than that of OPC, while long-term strength is similar. Applications include the following.

1. To permit increased speed in construction – for example, subsequent lifts of concrete can proceed more rapidly on account of its increased early strength.
2. In frosty weather there is less risk of the concrete freezing, since the cement has a higher early output.
3. Concrete made with rapid-hardening cement can be safely exposed to frost sooner, since it matures more quickly.

The term 'rapid-hardening Portland cement' is no longer used in the current edition of BS 12 though it is still used by manufacturers to refer to cements of strength class 52.5 N as defined in the standard.

## Sulphate-resisting Portland cement (BS 4027)

This cement (SRPC) has better resistance to sulphate attack than OPC (see 'Sulphate attack', page 66). The percentage of the sulphate-susceptible tricalcium aluminate is limited to 3.5 per cent in BS 4027 in order to minimise chemical combination with sulphates in solution. Sulphate-resisting cement may be produced by the addition of extra iron oxide before firing; this combines with alumina which would otherwise form $C_3A$, instead forming $C_4AF$ which is not affected by sulphates. Hence, sulphate-resisting cement may be slightly darker in colour than OPC. Applications include foundations in sulphate-bearing soils, in mortar for flues in which sulphur may be present from fumes, and in marine structures, since sea water contains sulphates. The current standard refers to cement strength grades as in BS 12 and to a low-alkali option of SRPC for use where there is a risk of alkali–silica reaction.

## Low-heat Portland cement (BS 1370)

The heat output of this cement (LHPC) is limited by BS 1370 to 250 J/g at the age of 7 days and 290 J/g at the age of 28 days. These may be compared with typical heat outputs of 330 and 400 J/g at the same ages for OPC. The reduction in heat output is obtained by means of lower quantities of the rapidly hydrating compounds $C_3S$ and $C_3A$. In order to produce satisfactory development of strength with time, the fineness of this cement is higher than that of OPC – it must not be less than $275 \, m^2/kg$. Although early strength is slightly less than that of OPC, long-term strength is similar.

The chief use of LHPC is in mass concrete, since there is a tendency for the heat to build up internally in such structures, causing cracking due to the temperature differential between inner and outer layers. Typical applications include large raft foundations and dams.

## Portland blast-furnace cement (BS 146)

This cement (PBFC) comprises a mixture of OPC and ground blast-furnace slag, the proportion of the latter not exceeding 65 per cent of the total. The slag contains mainly lime, silica and alumina (in order of decreasing amounts), which exhibit hydraulic action in the presence of calcium hydroxide liberated by the Portland cement on addition of water. Early strength is lower than that of OPC but PBFC generates less heat than OPC and produces better sulphate resistance. Ultimate strength is similar to that of OPC concrete and PBFC may be competitive in cost in regions in which the slag is produced. Ground blast-furnace slag may also be used in part replacement of cement, reducing cost and modifying properties as above. Cement grades as in BS 12 are specified.

## Pulverised fuel ash (PFA) in cements

PFA is a by-product of coal-powered power stations and is an example of a pozzolanic material – one which in the presence of lime (liberated by Portland cement) has hydraulic (cementing) properties. It can be incorporated in cement during manufacture (BS 6588; BS 6610) or used as a replacement at the time of batching. The DoE method of mix design now incorporates a section dealing with the design of such mixes.

## Microsilica

This is a by-product of electric arc furnaces used to produce silicon and ferro-silicon, most material being imported from Scandinavia. It is strongly pozzolanic and shows great potential as a replacement material for cement, improving resistance to sulphate attack and alkali–silica reaction.

## White Portland cement

This contains not more than 1 per cent of iron oxide, which is responsible for the grey colour of OPC. China clay, which is almost iron-free, is used instead of ordinary clay. Firing and grinding are also modified to prevent coloured matter being introduced. In consequence, the cement costs approximately twice as much as OPC. Setting is similar to OPC but, being composed mainly of $C_3S$, strength development may be rather faster than that of OPC (though within the limits of BS 12). The cement is used to produce white or coloured concretes – for the latter, pigments would be incorporated. These concretes may be employed for their aesthetic or light-reflecting qualities.

## Trends in composition/fineness of Portland cement

There is strong commercial competition between cement manufacturers who seek to supply a product resulting in maximum performance at the normal strength specification age of 28 days. In consequence, there has been a trend over the last 40 years or so towards finer cements with a higher $C_3S$ content. A concrete mix of a given specification will have a higher strength than previously since the cement is relatively fully hydrated at 28 days. Hence, to obtain a given strength, less cement will be needed making the mix more economical. There has been concern that, since there will be relatively little hydration after 28 days, subsequent strength gains will be reduced compared with 'traditional' cements, with possible implications for the longer term durability of the product. In consequence BS12 now specifies an **upper** strength limit for grade 42.5 cements to guard against the effects of very finely ground cements. A cement known as 'controlled fineness Portland cement' is also marketed although at present its use is mainly for some types of precast product where the coarser cement permits easier dewatering after compaction. Where there is concern about the long-term performance of the concrete a minimum cement content could be specified.

### British Standard tests

### British Standard test for setting time

In this test, a sample of cement is mixed with water to form a paste of standard consistency; checked by using a cylindrical plunger in the apparatus of Fig. 2.1. A 40 mm thick sample of this paste is then subjected to two types of penetration test, using Vicat apparatus. In the first, the penetration of a 1.13 mm diameter round needle is measured (Fig. 2.1). When this is between 4 and 6 mm from the base, the cement is said to have reached its 'initial set'. This would be related, for instance, to the time available for placing the concrete and must be not less than one hour for cements of grades 32.5 and 42.5 and not less than 45 minutes for cements of grades 52.5 and 62.5 (BS 12).

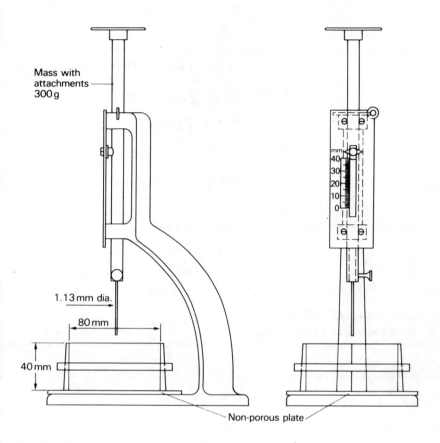

Fig. 2.1 Vicat apparatus fitted with initial set needle.

At a much later stage, a 1.13 mm diameter round needle projecting 0.5 mm from a small brass cylinder is applied (Fig. 2.2) and when the brass cylinder fails to mark the cement paste, the 'final set' has occurred. This would be related to the time after which concrete could be treated as 'solid', though BS 12 no longer gives requirements for the final setting time. It should be emphasised that both 'initial' and 'final' sets are arbitrarily selected points on a smooth and continuous curve relating stiffness to time.

## British Standard test for soundness

This test is primarily designed to detect the presence of any free lime which might be present in cement clinker. Such lime would, if present, gradually

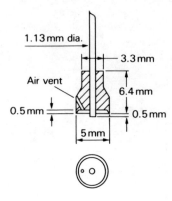

Fig 2.2    Enlarged view of final set needle for Vicat test.

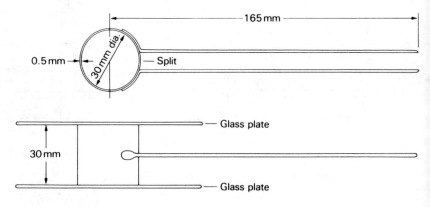

Fig. 2.3    Le Châtelier apparatus for measuring soundness of cement.

become exposed on hydration of the cement and, since it expands on 'slaking' with water, would cause overall expansion (unsoundness) in the hardened concrete with consequent risk of damage. Since the exposure of the lime takes some months to occur, an accelerated test is used to detect its presence. The Le Châtelier apparatus (Fig. 2.3) is used, consisting of a small brass cylinder containing a split, on each side of which are fixed long pointers to magnify any movement. A sample of cement mortar is allowed to harden in the apparatus and is then boiled for one hour. The expansion is noted after cooling and should not be greater than 10 mm. If the cement fails to comply, a sample is aerated by exposure to air, of humidity between 50 and 80 per cent, for 7 days. This causes slaking of any lime to which air has ready access. The expansion of the cement on repeating the soundness test should not exceed 10 mm.

## British Standard test for strength

The current BS 12 test for compressive strength of cements (EN 196, 1987) relates to samples of size $40 \times 40 \times 160$ mm, made from a 3:1 sand/cement mortar with water/cement ratio 1:2. The sample is first tested in flexure and then the broken pieces are tested in compression using a special $40 \times 40$ mm jig. The test employs a standard CEN sand of specified grading, the mortar being prepared by mechanical mixing. Standard sands such as Leighton Buzzard sand should be checked against the EN 196 specification before being used. BS 12 (1991) requirements for the compressive strength of mortar cubes are given in Table 2.2.

## The structure of hydrated cement

The most important components of Portland cement from the strength development point of view are $C_2S$ and $C_3S$ which, on hydration, form the same compounds in differing proportions:

$$2(2CaOSiO_2) + 4H_2O \rightarrow 3CaO.2SiO_2.3H_2O + Ca(OH)_2$$

$$2(3CaOSiO_2) + 4H_2O \rightarrow 3CaO.2SiO_2.3H_2O + 3Ca(OH)_2$$

$$\underset{\text{Calcium}}{\phantom{x}} \underset{\text{silicate}}{\phantom{x}} \underset{\text{hydrate}}{\phantom{x}} \qquad \underset{\text{Calcium}}{\phantom{x}} \underset{\text{hydroxide}}{\phantom{x}}$$

(Note that the full formulae for $C_2S$ and $C_3S$ have been given here to enable the equations to be balanced.)

The *physical* form of these compounds plays a most important part in the behaviour of concrete. The calcium silicate hydrate forms extremely small

Table 2.2  BS 12 (1991) requirements for the strength of mortar cubes

| | Strength requirement ($N/mm^2$) | |
| --- | --- | --- |
| *Age* | *Cement (grade 42.5 N)* | *Cement (grade 52.5 N)* |
| 2 days | ≥10 | ≥20 |
| 28 days | ≥42.5  ≤62.5 | ≥52.5 |

fibrous, platey or tubular crystals, which can be regarded as a sort of rigid sponge, referred to as cement 'gel'. This gel must be saturated with water if hydration is to continue. The calcium hydroxide forms much larger platey crystals. These dissolve in water, providing hydroxyl ($OH^-$) ions, which are important for the protection of steel in concrete. As hydration proceeds, the two crystal types become more heavily interlocked, increasing the strength of the concrete, though the main cementing action is provided by the gel which occupies two-thirds of the total mass of hydrate.

## High-alumina cement

It is important to appreciate that high-alumina cement (HAC) is not a Portland cement and has quite different properties. It was developed in the 1930s to resist sulphates and is manufactured from limestone and bauxite (aluminium ore). The main cementing ingredient is monocalcium aluminate (CA using cement notation) and since $C_3A$ is not present, the cement normally has good sulphate resistance. It also exhibits very rapid-hardening properties (but is not quick setting). In consequence, the cement has been widely used for precasting, since the turn-round time of expensive steel moulds is reduced. The design of HAC concrete is in fact based on *24 hour* strength rather than 28 day strength for OPC concrete. Concretes made with HAC are usually a darker colour than Portland cement equivalents, reflecting the dark grey colour of the cement. Use of HAC concrete for structural purposes in building has been banned since 1973 following the collapse of a school roof and swimming pool roof. At least partly responsible for the collapse was the phenomenon of 'conversion' in which a rearrangement of the crystalline hydrate results in the formation of pores in the concrete, leading to permeability and reduced strength. Conversion was initially thought to be associated with high temperatures and humidities (as might occur in a swimming pool) though it is now known to occur slowly in normal conditions. The strength reduction can be minimised by use of low water/cement ratio concretes. These concretes not only have higher initial strength, but on conversion it is possible that residual unhydrated cement can hydrate, filling pores as they are formed. There are many HAC structures still in existence; these should be periodically checked to determine the degree of conversion and the consequent effect on strength. On heating to high temperatures HAC forms a ceramic bond and is hence widely used for refractory purposes such as flue linings.

## Aggregates for concrete

Aggregates are used in concrete for the following reasons:

- They greatly reduce cost.

- They reduce the heat output per unit volume of concrete and hence reduce thermal stress.
- They reduce the shrinkage of the concrete.
- They help produce a concrete with satisfactory plastic properties.

There may be other reasons for which 'special' aggregates would be employed; for example:

- Low-density concrete; to decrease foundation loads, increase thermal insulation and reduce thermal inertia (lightweight aggregates).
- High-density concrete; for example, as required for radiation shielding (barytes (barium sulphate) or iron-based aggregates).
- Abrasion-resistant concretes for floors (granite or carborundum aggregates).
- Improved fire resistance (limestone, lightweight aggregates such as expanded pulverised fuel ash).
- Decorative aggregates; for example, crushed granite is available in several different colours which can be revealed by use of an exposed aggregate finish.

## Maximum aggregate size

As a general rule, the maximum size of an aggregate should be as large as possible, since larger aggregates result in a lower sand requirement, hence, a lower specific surface of the aggregate overall and consequently reduced water and cement requirements. The resulting concrete should therefore be more economical and exhibit lower shrinkage for a given strength. There are, of course, constraints on the maximum size that can be employed. First, the maximum aggregate size should not exceed approximately one-fifth of the minimum dimension in the structure, consideration also being given to spacings between reinforcing bars. Secondly, the aggregate size chosen will need to be a size that is readily available. The standard sizes, together with typical applications, are given in Table 2.3.

Table 2.3 British Standard maximum aggregate sizes with typical applications.

| Nominal max. size (mm) | Application |
| --- | --- |
| 40 | Mass concrete, road construction |
| 20 | General concrete work, including reinforced and prestressed concrete |
| 10 | Thin sections, screeds over 50 mm thickness |
| 5 | Screeds of 50 mm thickness or less |

Aggregates which are largely retained on a 5 mm mesh sieve are described as coarse aggregates. These may comprise uncrushed natural gravels which result from the natural disintegration of rock or crushed stones and crushed gravel, produced by crushing hard stone and gravel respectively. Aggregates that largely pass a 5 mm sieve are referred to as sands. (The current edition of BS 882, 1992, uses the term 'sand' in preference to 'fine aggregate'. The term 'fines' is reserved for material such as silt or clay, passing the 75 μm sieve.) Sands may comprise natural sands – that is, sands resulting from natural disintegration of rock or crushed stone – or crushed gravel sands.

## Grading of aggregates

It is customary for aggregates for concrete to be *continuously graded* from their maximum size down to the size of the cement grains, since this ensures that all voids between larger particles are filled without an excess of fine material. (It will be recalled that an excess of fine material results in increased water and cement requirements, as explained under the heading of 'Maximum aggregate size'.) This effect is illustrated in Fig. 2.4.

Both crushed and uncrushed natural gravels have naturally continuous gradings, but to comply with BS 882 the gradings must be within certain limits. Gradings are determined by passing aggregates through a set of standard sieves, the particular sieves used depending on the maximum aggregate size (Experiment 2.1). The grading is then defined by the percentage of the total sample used which passes each sieve. Table 2.4 shows the sieves used for concreting sands, together with specimen results.

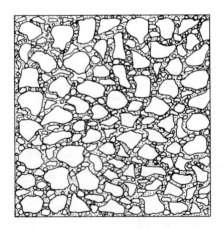

Fig. 2.4 Simple illustration of the use of graded aggregate in concrete. Spaces between larger particles are filled by progressively smaller particles.

Table 2.4 Example of a sieve analysis for a sand for concrete

| Sieve size | Mass retained (g) | Cumulative mass retained (g) | Mass passing (g) | Percentage passing |
|---|---|---|---|---|
| 10 mm | 0 | 0 | 287 | 100 |
| 5 | 6 | 6 | 281 | 98 |
| 2.36 | 17 | 23 | 264 | 92 |
| 1.18 | 32 | 55 | 232 | 81 |
| 600 μm | 48 | 103 | 184 | 64 |
| 300 | 81 | 184 | 103 | 36 |
| 150 | 86 | 270 | 17 | 6 |
| Passing 150 | 17 | 287 | — | — |

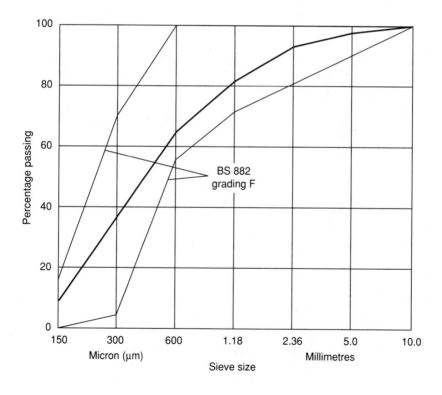

Fig. 2.5 Sand grading corresponding to sieve analysis results given in Table 2.4. The F grading envelope into which this aggregate fits is also indicated.

These are plotted in Fig. 2.5. Note that the sieve sizes are not plotted to a linear scale, successive sizes being spaced equally instead. Points are joined by straight lines. Figure 2.5 also includes the limits of BS 882, grading F, into which this particular sand fits. The current edition of the standard gives three gradings: C, M and F, corresponding to coarse, medium and fine sands, replacing the four 'zones' of earlier editions. This reflects the fact that the previous zones were rather restrictive in terms of sand acceptance; materials which did not comply with a 'zone' were often perfectly acceptable for use in concrete, with suitable mix proportioning. There are no rigid distinctions between the C, M and F gradings and an aggregate may comply with two of the envelopes. The aggregate shown in Fig. 2.5, for example, complies with grading M as well as with grading F. However, very fine gradings may cause problems and BS 882 states that sands should comply with gradings C or M for heavy duty concrete floor finishes.

Coarse aggregates are not assigned grading 'zones', since the range of particle size is, in general, smaller than that of sands. They should, however, fit into a BS grading envelope for that particular size and type – see Experiment 2.1.

Table 2.5 indicates that there are three ways in which a continuously graded aggregate of maximum size 20 mm may be stockpiled and batched. The simplest method would be to have a single stockpile of 20 mm 'all in' aggregate, sometimes referred to as 'ballast', since this material is relatively cheap and only a single quantity then needs to be batched. This method gives, however, only poor control over the distribution of particle sizes within each batch quantity – coarse particles are prone to separation from fine material, especially if stockpiling technique is poor. Successive batches of the resulting concrete would therefore tend to have variable water requirements for a given workability, depending on the fineness of individual batches of aggregate. While this method of batching is satisfactory for lightly stressed concrete – such as foundations and ground floor slabs in domestic dwellings – it would not be recommended for structural quality concrete. The second alternative is to use '20 mm graded coarse aggregate' and sand,

Table 2.5  Three alternative ways of batching graded aggregate of maximum size 20 mm

| No. of Stockpiles | Aggregate size | | | |
|---|---|---|---|---|
| | 20 mm | 10 mm | 5 mm | 150 μm |
| 1 | 20 mm 'all in' aggregate (ballast) | | | |
| | ←——————————————————————————————————→ | | | |
| 2 | 20 mm graded coarse aggregate | | sand | |
| | ←——————————————————→◄—→←——————————————→ | | | |
| 3 | 20 mm single-size coarse aggregate | 10 mm single-size coarse aggregate | sand | |
| | ←————————————→◄—→←——————————→◄—→←————————→ | | | |

thereby ensuring that the ratio of quantities of aggregate of sizes above and below 5 mm is maintained accurately constant. This 'two stockpile' method is widely used in concrete production. Even closer control results when three stockpiles are used:

- 20 mm single-sized coarse aggregate
- 10 mm single-sized coarse aggregate
- sand

This permits very close control over the coarser sizes which are most prone to segregation. The extra cost of material and additional batching may, in high-quality work, be offset by the reduced variability of the resultant concrete.

## Aggregate density

Before classifying aggregates into density groups, it is necessary first to examine the meaning of the term 'density', since a number of differing definitions exist, according to the context of use.

### Relative and solid density

Most aggregates comprise a mass of solid material containing air pores which may or may not be accessible to water. The term which describes the density of the solid material itself is 'relative density' (formerly specific gravity) or density relative to that of water, though several definitions are possible according to the way in which accessible pores are treated. Perhaps the most important of these is relative density in the 'saturated surface dry' (SSD) state – that is when all accessible pores are full of water but the aggregate surface is dry. This relative density usually forms the basis of mix design methods and is defined as:

$$\text{Relative density (saturated surface dry)} = \frac{\text{Mass of given sample of SSD aggregate particles, including absorbed water}}{\text{Volume of water displaced by saturated surface dry sample} \times \text{density of water}}$$

Note that because it is a ratio, relative density has no units.

Relative density could be defined in other ways – for example, by taking the mass of *dry* aggregate with the same volume as given above. (This is referred to as the 'oven dry' relative density.) However, in the case of most natural aggregates, void contents are small, so that differences between the various definitions of relative density are correspondingly small. The term 'solid density' will be taken to mean $1000 \times$ relative density (say, on a saturated surface dry basis), normally having the units $kg/m^3$.

### Bulking and bulk density

When aggregates are loosely packed together or stockpiled, large volumes of air are trapped between particles – usually many times the volume of air present *within* particles. This is referred to as 'bulking' and for coarse aggregates it amounts to between 30 and 50 per cent of the total space occupied. The extent of bulking of sands depends very much on their moisture content; the void content of dry sand may be quite small – say, 20 per cent. This can increase to 40 per cent when 5–10 per cent moisture is present, thereafter decreasing as further moisture tends to assist particles to consolidate. The bulking at intermediate moisture contents is the result of thin water films increasing friction between sand particles. The situation is illustrated in Fig. 2.6 (see Experiment 2.2).

Bulking of aggregates produces uncertainty in the solid content of aggregates batched by volume and for this reason batching by weight (mass) is much preferred, except for concrete having only nominal performance requirements. Hence, most concrete production equipment is geared to weight batching.

*Low-, medium- and high-density aggregates*
Most natural aggregates have solid densities within quite a narrow range of values – between 2400 and 2700 kg/m$^3$ (corresponding to relative densities of between 2.4 and 2.7). These would result in concrete having densities normally in the range 2200–2500 kg/m$^3$ – slightly lower than the corresponding aggregate densities on account of the water content of the concrete.

Widely used in the construction industry are lightweight aggregates, which comprise highly porous particles. Examples of synthetic lightweight aggregates are sintered pulverised fuel ash, expanded clay, blast-furnace slag and expanded slate. Natural lightweight aggregates such as pumice may also be used. Densities as low as 500 kg/m$^3$ can be obtained in this way (see page 60 'Lightweight concretes').

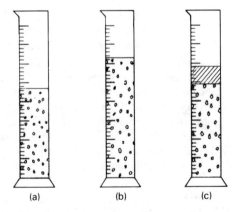

Fig. 2.6   Volume occupied by a given mass of sand when: (a) dry; (b) damp; (c) wet.

There are certain situations in which high-density concretes are required – for example, for radiation shielding for nuclear reactor vessels. Such concretes are produced by use of high-density aggregates such as barytes (barium sulphate), relative density 4.1, or magnetite (iron ore), relative density 4.5, or even metallic aggregates such as steel shot which may have densities over 7.0, resulting in concrete with over twice the normal density.

## Silt and clay (defined in BS 882 as 'fines')

These may be defined as material passing a 75 µm sieve (compared with the smallest BS sieve size for fine aggregate, 150 µm). They are harmful to concrete if present in substantial amounts, since they increase the specific surface (see page 16) and hence the water requirement of a mix, resulting in lower strength unless the cement content is also increased.

Sands are likely to contain more silt than coarse aggregates and a rapid estimate of the silt/clay content of sands (other than crushed stone sands) can be made by a 'field settling test'. The test involves shaking a measuring cylinder containing a mixture of a 1 per cent salt and sand solution (Experiment 2.3). The salt helps to separate the silt into a separate layer whose volume can be measured after being allowed to stand for three hours (Fig. 2.7). If the amount is more than approximately 8 per cent by volume, separate tests are necessary to determine whether the material passing the 75 µm sieve *by weight* is in excess of the BS 882 limit of 4 per cent for natural sand and crushed gravel sand or 16 per cent for crushed rock sand.

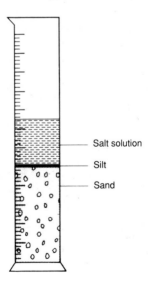

Salt solution

Silt

Sand

Fig. 2.7  Field-settling test for detection of silt/clay in sand.

Coarse aggregates can be checked for silt and clay by visual inspection. When in doubt, the test of BS 812, Part 103.1, should be used to ascertain whether the silt content is in excess of BS 882 limits.

## Moisture content

It is advisable during quality concreting to check moisture contents of aggregates prior to batching, so that accurate adjustment of batch weights can be made as necessary; batch quantities of aggregates must be increased if free water is present and the amount of free water in aggregates should be deducted from the water to be added. (A further possibility, though this is not common with natural aggregates in the UK, is that extra water may be required if the aggregates are dry, such that they *absorb* water on wetting. The latter situation is quite common with lightweight aggregates which absorb substantial quantities of water.) Sands normally trap more water than coarse aggregates, since there are more points of contact between particles.

The principles of three types of test are given below and in each case the aim will be to obtain the *free* water content of the aggregates since no adjustment is necessary in mixing water for moisture which is absorbed *within* aggregate particles.

### Drying methods

In these a sample of damp aggregate is carefully weighed ($M_w$) and then dried until it reaches a saturated surface dry condition. A suitable method is by hairdryer, sand being dried uniformly until it is just free flowing. Coarse aggregates should be dried until all traces of surface moisture have disappeared. A frying pan on a gas or electric ring may alternatively be used (though care is necessary to ensure uniform heating of the sample). The aggregate is then weighed again ($M_d$). The moisture based on wet weight is then given by:

$$\text{Free moisture content (per cent)} = \frac{M_w - M_d}{M_w} \times 100$$

### Siphon can

This relies on the fact that the volume of an accurately measured mass of aggregate of given density depends on the amount of moisture in it, moisture having a density of only approximately 40 per cent that of most natural aggregates (see Experiment 2.4). Hence the greater the moisture content, the greater the volume of a given mass (normally 2 kg). A sample of dried aggregate is necessary for the test and, to give free water content, this should be in the saturated surface dry condition. Accurate weighing of dry and damp samples is necessary in this method, which is suitable for both coarse aggregate and sand.

*Calcium carbide method*
A rapid estimate of the moisture content of sand can be made by mixing a standard quantity of aggregate with an excess of calcium carbide in a pressure vessel (see Experiment 2.5). The latter reacts with moisture, releasing a gas whose pressure is measured. The pressure scale is calibrated to read moisture content based on wet weight directly. This method would normally measure the *free* moisture contents of hard materials such as aggregate, since the calcium carbide cannot easily penetrate pores in the stones. The sample size for this test is normally very small so that careful sampling is necessary for sand and the test is not suitable for coarse aggregate – there would be too few particles to give a representative sample.

## Correction for moisture content

A method of correction of batch quantities for a given volume of concrete will be illustrated by means of a worked example.
Determine the correct batch quantities for 1 cubic metre of concrete if the quantities based on saturated surface dry aggregates are:

| | |
|---|---|
| Cement | 310 kg |
| Sand | 650 kg |
| Coarse aggregate | 1190 kg |
| Water | 180 litres |

Moisture contents of sand and coarse aggregates are 4.5 and 1.5 per cent, respectively, based on wet weight.

*Answer*
The moisture in 650 kg of sand equals

$$\frac{4.5 \times 650}{100} = 29.25 \text{ kg}$$

Hence *add* this amount to the quantity of sand giving

$$650 + 29.25 = 679.25 \text{ (say 680) kg}$$

as the correct batch mass of sand.
The moisture in 1190 kg of coarse aggregate equals

$$\frac{1.5 \times 1190}{100} = 17.85 \text{ kg}$$

Again this amount must be added to the quantity of coarse aggregate, giving

$$1190 + 17.85 = 1207.85 \text{ (say 1208) kg}$$

as the correct batch quantity of coarse aggregate. To obtain the correct batch quantity of water, the water contents above are subtracted from the given batch quantity:

$$180 - (29.25 + 17.85) = 132.9 \text{ (say 133) kg, or litres}$$

It may be argued that the above method is inaccurate, since the extra material added to sand and coarse aggregates also contains water, which should be allowed for. Strictly speaking this is true, but the errors in percentage terms are small enough to be ignored and the actual errors would normally be smaller than the tolerances in batching equipment. The above method is given on account of its simplicity.

## The design of concrete mixes

The object of mix design is to produce, as economically as possible – by systematic analysis of materials' properties and knowledge of how they affect concrete properties – batch quantities for a given volume of concrete such that the properties of both the fresh and hardened material are as required for the specific purpose.

Consideration has already been given to properties of cements and aggregates and it is now necessary to identify the important properties of the concrete itself in the fresh and hardened states and to examine the effect of materials types and proportions on them.

Two important terms, *water/cement ratio* and *aggregate/cement ratio*, are often used in the context of mix design and are therefore now defined:

$$\text{Water/cement ratio} = \frac{\text{Mass of water in a concrete sample}}{\text{Mass of the concrete sample}}$$

Note that *free* water/cement ratio is the most important since it is this that affects strength and durability. Free water /cement ratio would be based on the free water content of the mix – that is, water absorbed in the aggregates is disregarded.

$$\text{Aggregate/cement ratio} = \frac{\text{Mass of aggregate in concrete sample}}{\text{Mass of cement in that sample}}$$

For most concrete mixes the aggregate/cement ratio would be in the range 4–10, small ratios indicating rich, expensive mixes and high ratios indicating lean, cheaper mixes. Although mortars are still commonly specified in terms of aggregate (sand)/cement ratios, the richness of concrete mixes is now more commonly specified in terms of cement content per cubic metre. For example, an aggregate/cement ratio of 4 would correspond to about 450 kg of cement per cubic metre while an aggregate/cement ratio of 10 would correspond to about 200 kg of cement per cubic metre.

## Properties of fresh concrete

The fresh concrete must be satisfactory in relation to its workability and its cohesion.

# Workability

This is the most important term relating to fresh (plastic) concrete. Workability may be defined as that property of the concrete which determines its ability to be placed, compacted and finished. Of these three operations, the greatest emphasis should be placed on compaction (elimination of air voids), since the consequences of inadequate compaction are serious. Workability may be measured by the slump test, compacting factor test and Vebe test (BS 1881, Parts 102–104), brief details of each now being given. For each test it is essential that a representative sample of concrete be obtained. Details of how to obtain such samples are given in BS 1881, Part 101.

In the slump test (Experiment 2.6) concrete is placed into a special cone in three layers of equal height, each layer being compacted in a standard manner. On removing the cone the concrete 'slumps', the slump being equal to the difference between the height of the cone and the highest point on the concrete (Fig. 2.8). When the concrete slumps evenly, a 'true' slump is said to occur. In less cohesive concretes, collapse may occur on an inclined shear plane, producing a 'shear' slump. A wet mix of low cohesion may produce a 'collapse' slump. Slumps may vary from zero for dry concrete mixes, through 50 to 70 mm for medium workability concretes, to 150 mm or more for wet or harsh mixes. The test is widely used as an on-site check, though it is not suitable for dry mixes which tend to give low or zero slump.

In the compacting factor test (Experiment 2.7) concrete is gently loaded into a hopper (Fig. 2.9), then allowed to fall vertically into a similar lower hopper and then into a cylinder by the action of gravity. The concrete in the cylinder is then struck off level and weighed. The cylinder is then emptied and refilled, this time fully compacting the concrete. The 'compacting factor' is:

$$\frac{\text{Mass of partially compacted concrete}}{\text{Mass of fully compacted concrete}}$$

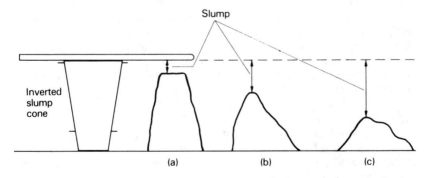

Fig. 2.8   Method of measuring slump in slump test together with types of slump: (a) true slump; (b) shear slump; (c) collapse slump.

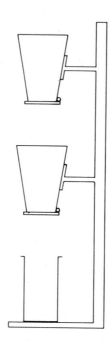

Fig. 2.9   Compacting factor apparatus.

The value will approach unity for wet mixes (for which the test is not suitable) and may be as low as 0.65 for very dry mixes.

The third test, the Vebe test (Experiment 2.8) utilises a slump cone in a cylindrical container fixed to a small vibrating table (Fig. 2.10). A slump test is first carried out and then a Perspex disc attached to a vertical guide is allowed to rest gently on the concrete. The table is then made to vibrate and the time taken for the underside of the Perspex disc to become completely covered in concrete is measured in seconds. The time may vary from a few seconds for wet mixes (for which the test is not suitable) to over 20 seconds for dry concretes.

*Comments on workability tests*
From the descriptions given above it will be evident that the slump test is the most suitable test for the measurement of medium to high workabilities. The test has the advantage that the equipment is simple, portable and does not require an electricity supply.

The compacting factor test is sensitive to medium and low workabilities,

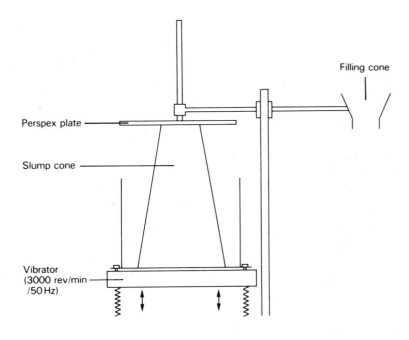

Filling cone

Perspex plate

Slump cone

Vibrator
(3000 rev/min
/50 Hz)

Fig. 2.10   Vebe consistometer.

though its use is not included in the current DoE design document *Design of normal concrete mixes* since 'it is not possible to establish consistent relationships between it and the slump or Vebe tests'. The compacting factor test may nevertheless be useful as a control test for drier concretes and is used in civil engineering applications.

The Vebe test has the disadvantage of requiring mains electricity, though it is very sensitive to changes in the water content of dry mixes, this providing the main basis for its use.

It is important to appreciate that all the above tests are used to check the workability of the concrete, once mixed; and that, since design methods rarely give a very accurate water batch quantity, even with moisture content correction, there is a need to be able to judge the workability of the concrete while in the mixer. The mixer operator may well be able to do this, for example by the appearance of the concrete or the tone of the engine or motor which is used to drive the mixer, since drier mixes may increase the power consumption. Hence an experienced mixer operator is a very useful part of the construction team in producing concrete of consistently correct workability.

*Workability requirements*
The workability of concrete for a given situation should be the minimum value that will ensure full compaction with the plant available, due consideration being given to difficulties which may arise resulting from problems such as access, congested reinforcement and depth of pour. Table 2.6 gives the main workability categories and applications.

*Factors affecting workability*
The chief factor affecting workability is water content (usually expressed as volume in litres per cubic metre of concrete). For a given aggregate type and size, workability is highly sensitive to changes of water content – for example, in the case of 20 mm uncrushed aggregate, a water content of 160 litres per cubic metre would result in low workability, while a water content of 190 litres would result in high workability.

Also affecting workability are maximum aggregate size and shape. Since a smaller maximum aggregate results in a higher specific surface (surface area per unit mass) of the mix overall, more water is required to 'wet' the larger surface areas involved. For a given aggregate type, decreasing the maximum size from 20 to 10 mm would increase the water requirement by approximately 10 per cent. Increasing the maximum aggregate size from 20

Table 2.6   Workability categories and applications

| Workability category | Slump (mm) | Compacting factor | Vebe time (seconds) | Applications |
|---|---|---|---|---|
| Extremely low | 0 | 0.65–0.7 | Over 20 | Lean mix concrete for roads (compacted by vibrating roller). Precast paving slabs |
| Very low | 0–10 | 0.7–0.75 | 12–20 | Roads compacted by power operated machines |
| Low | 10–30 | 0.75–0.85 | 6–12 | High-quality structural concrete. Mass concrete compacted by vibration |
| Medium | 30–60 | 0.85–0.95 | 3–6 | Normal purposes-reinforced concrete compacted by vibrating poker or manually |
| High | 60–180 | 0.95–1.0 | 0–3 | Areas with congested reinforcement, concrete for placing underwater |

to 40 mm would naturally have the reverse effect. Crushed (angular) aggregates similarly result in a larger specific surface than uncrushed (irregular) aggregates and, therefore, an increase of 15–20 per cent in water requirement. Aggregates having a coarse surface texure will also increase water requirements, though it should perhaps be added that, for high-strength concretes, angular shapes combined with a rough surface texture allow fuller exploitation of a hard aggregate material. The strength of concretes made with natural flint-based gravels is, for example, limited at the high strength end by a tendency of the smooth aggregate surface to debond from the cement mortar under stress.

Although use of the terms 'water/cement ratio' and 'aggregate/cement ratio' in the context of workablity is limited, it may be said that:

- If the water content of a mix is increased, other quantities remaining constant, both water/cement ratio and workability would increase – hence an indirect relationship exists between water/cement ratio and workability.
- If the water content of a mix is increased, keeping water/cement ratio constant, more cement and hence less aggregate would be present in a given volume; hence, increasing the water in this way would increase the workability and decrease the aggregate/cement ratio and, again, an indirect relationship is seen to exist between aggregate/cement ratio and workability.

## Cohesion

Though perhaps not of such critical importance as workability, the choice of concrete of correct cohesion is an essential part of the design process. Cohesion (or cohesiveness) may be defined as the ability of the fresh concrete to resist segregation, though a cohesive mix will also be easier to finish than a relatively harsh mix. The cohesion in a concrete mix is provided by the fine material, that is, the finer fractions of sand and the cement. It follows that richer mixes which have a higher cement content should, in general, have a reduced sand requirement and this is obtained when concrete is designed by the DoE method of mix design. It should perhaps also be stated that cohesion should be the minimum required for the purpose since, if, in common with workability, cohesion is higher than necessary, the water content of the mix will also be unnecessarily high, leading to the disadvatages described earlier.

## Properties of hardened concrete

In the context of mix design the most important properties of concrete are strength and durability.

## Strength

This is normally considered to be the most important property in relation to mature concrete. In the UK, 'strength' most commonly means compressive strength as measured by cubes manufactured, cured and tested according to BS 1881 (see Experiment 2.9). The strength of concrete is affected by the following aspects of mix materials and proportions:

- *Free water/cement ratio.* As the quantity of free water in a mix increases in relation to the quantity of cement, the density and, consequently, the strength of the concrete decrease. At very low water/cement ratios (e.g. below 0.4) a significant part of the cement never hydrates, due to inadequate space and/or inadequate water. Since unhydrated cement is of high strength, the resulting product will also be of high strength, provided full compaction is achieved. At water/cement ratios above about 0.4, the expansion of cement on hydration is insufficient to occupy the space previously filled with water. Hence porosity increases and strength decreases. At water/cement ratios in the range 0.4–1.0, strength increases progressively with decreasing water/cement ratio. At values below 0.4 the strength of fully compacted concrete continues to rise, reaching values over $100 \, \text{N/mm}^2$ at 0.2 or less, though the concrete is very difficult to compact at such values.
- *Aggregate properties.* Crushed aggregates generally result in higher strength than uncrushed aggregates, since they form a better key with hydrated cement. To obtain very high strengths the use of crushed aggregates may be essential.
- *Cement type.* Where high early strength is required, the use of rapid-hardening Portland cement may be considered, though long-term strength will be similar to that of OPC.

## Durability

The durability of normal concrete depends mainly on the permeability, and hence on the porosity, of the hydrated cement. Hence the arguments are similar to those given under strength, with low water/cement ratios resulting in greatest durability. To illustrate this, concretes of free water/cement ratio 0.4, 0.6 and 0.8 would result in very good, moderate and poor durability respectively in concrete subject to average exposure. It is not always easy to measure free water/cement ratio in production, so that durability is sometimes specified alternatively by means of minimum cement content on the basis that, for a given workability, the water content would be constant; hence, by specifying a cement content, one is effectively specifying a water/cement ratio. Another way of specifying durability indirectly is by specifying strength, though the correlation depends on aggregate properties.

## DoE method of mix design

The description given here is intended to illustrate the principles of the method; the student would be well advised to carry out a number of design exercises in order to become familiar with the design procedure. The DoE publication (1988 edition) also includes information necessary to design concretes for indirect tensile strength, containing entrained air and using cement replacements. These are not given here, since they are easily followed by reference to the DoE publication once the basic method of design is understood.

The method is divided into sections which will be taken in turn. (The section numbers refer to the sections in the standard mix design form, Fig. 2.11. Subsection numbers in Fig. 2.11 are given in the margin.)

### Section 1. Strength consideration

*1.1* Before describing the method of designing for strength it is necessary to understand the meaning of the term ' characteristic strength' $(f_c)$, which is normally used in strength specifications. Concrete, in common with other materials, is inherently variable, due to within-batch variations of materials, between-batch variations, tolerances and possible errors in batching, mixing, compaction, testing and so on. Hence statistical terminology must be used and the term 'characteristic strength' implies a strength with a permissible percentage of failures. For example, a characteristic 28-day compressive strength of $20 \text{ N/mm}^2$ with 5 per cent failures would imply that 95 per cent of results need to be at, or higher than, the strength level of $20 \text{ N/mm}^2$. The first step in mix is to obtain a target mean strength $f_m$, which is the mean strength that would be required for the stated characteristic strength, $f_c$. Clearly, the target mean strength will need to be higher than the characteristic strength since *half* of the results will be expected to fall below the former. The increment which must be applied depends on two factors:

(a) The percentage failures allowable. If, for example, only 1 per cent failures were permitted below the stated characteristic strength, the target mean strength would need to be higher than if 5 per cent failures were allowed. These percentage failure rates are taken account of by $k$ factors. Table 2.7 gives common values.

*1.2* (b) The variability of the concrete. If, due to say poor quality control, the concrete is more variable, an extra margin will need to be built into the design to compensate. The variability of concrete strength is represented by its standard deviation $(s)$, which will be defined later. To allow for concrete variability, the $s$ value should be known and this, of course, requires that

42

**Concrete mix design form**

Job title...........................................................................................................................

| Stage | Item | Reference or calculation | Values |
|---|---|---|---|

**1**  1.1 Characteristic strength — Specified — _____ N/mm² at _____ days
Proportion defective _____ %

1.2 Standard deviation — Fig 2.12 — _____ N/mm² or no data _____ N/mm²

1.3 Margin — Table 2.7 — (k = _____ ) _____ × _____ = _____ N/mm²
_____ N/mm²

1.4 Target mean strength — _____ + _____ = _____ N/mm²

1.5 Cement type — Specified — OPC/SRPC/RHPC

1.6 Aggregate type: coarse — Crushed/uncrushed
Aggregate type: sand — Crushed/uncrushed

1.7 Free-water/cement ratio — Table 2.8 Fig 2.13 _____

1.8 *Maximum free-water/cement ratio* — *Specified* — _____ } Use the lower value _____

**2**  2.1 Slump or Vebe time — Specified — Slump _____ mm or Vebe time _____ s

2.2 Maximum aggregate size — Specified — _____ mm

2.3 Free-water content — Table 2.9 _____ kg/m³

**3**  3.1 Cement content — _____ ÷ _____ = _____ kg/m³

3.2 *Maximum cement content* — *Specified* — _____ kg/m³

3.3 *Minimum cement content* — *Specified* — _____ kg/m³
use 3.1 if ≤ 3.2
use 3.3 if > 3.1 — _____ kg/m³

3.4 Modified free-water/cement ratio — _____

**4**  4.1 Relative density of aggregate (SSD) — _____ known/assumed

4.2 Concrete density — Fig 2.14 — _____ kg/m³

4.3 Total aggregate content — _____ − _____ − _____ = _____ kg/m³

**5**  5.1 Grading of sand — Percentage passing 600μm sieve — _____ %

5.2 Proportion of sand — Fig 2.15 — _____ %

5.3 Sand content — _____ × _____ = _____ kg/m³

5.4 Coarse aggregate content — _____ − _____ = _____ kg/m³

| Quantities | Cement (kg) | Water (kg or L) | Sand (kg) | Coarse aggregate (kg) 10mm | 20mm | 40mm |
|---|---|---|---|---|---|---|
| per m³ (to nearest 5 kg) | | | | | | |
| per trial mix of ____ m³ | | | | | | |

Items in italics are optional limiting values that may be specified

OPC = ordinary Portland cement; SRPC = sulphate-resisting Portland cement; RHPC = rapid-hardening Portland cement. SSD = based on a saturated surface-dry basis.

Fig. 2.11   Concrete mix design form. (Crown copyright.)

Table 2.7   *k* factors used in statistical control

| Percentage | k |
|------------|------|
| 16 | 1.00 |
| 10 | 1.28 |
| 5 | 1.64 |
| 2 | 2.05 |
| 1 | 2.33 |

some cube test results be available (usually at least 40 for the *s* value to be reliable). Where concrete production is starting for the first time or a new type of mix is to be used, standard deviations cannot be known, since no results will be available; hence, in these circumstances an *s* value must be assumed and this is taken to be relatively high in order to 'play safe'. Figure 2.12 gives initial values suggested by the DoE method (graph A), together with minimum *s* values (graph B) in order to guard against the dangers of taking too small a value of *s*, once results are available. Note that, to some extent, *s* values reduce as characteristic strength decreases, since the variability of weaker concrete would be expected to reduce (negative strengths are impossible).

*1.3* The increment to be added to the characteristic strength $f_c$ is the product of the two terms described: *k* factor and the standard deviation *s* . Hence:

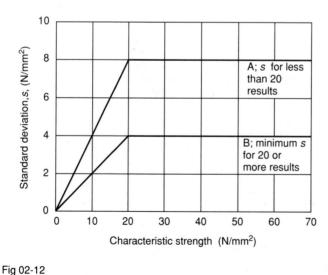

Fig 02-12

Fig. 2.12   Relationship between standard deviation and characteristic strength. (Crown copyright.)

*1.4*
$$f_{\mathrm{m}} \quad = \quad f_{\mathrm{c}} \quad + \quad ks$$
target mean strength     characteristic strength

The product *ks* is often referred to as the 'current margin'.

*1.5, 1.6*  It has already been explained that the strength of the concrete is affected by the free water/cement ratio, the aggregate type and the cement type (as well as age). Hence, the remaining part of this section involves determining the free water/cement ratio.

*1.7*  First, the strength appropriate to a free water/cement ratio of 0.5 and the materials specified is obtained from Table 2.8. This strength is then located on the graph of Fig. 2.13, using the 0.5 water/cement ratio starting line. Note that, since curves on this figure are separated by strength increments of 5 or $10\,\mathrm{N/mm^2}$ at the 0.5 free water/cement ratio line, they will not generally correspond exactly to the strengths given in Table 2.8. To find the free water/cement ratio corresponding to the *required* target mean strength, move *parallel* to the nearest two lines on Fig. 2.13 (unless the strength happens to fall on a line) from the point located towards higher or lower strength, as required, until the required target mean strength is obtained (reading horizontally). At this point, the correct free water/cement ratio is obtained by reading *vertically downwards*. Since this procedure may require some practice, it is worth checking the result for accuracy. The water/cement ratio obtained should be checked against any maximum value required for durability and the lower of the two values taken.

Table 2.8   Approximate compressive strengths ($N/mm^2$) of concrete mixes made with a free-water/cement ratio of 0.5. (Crown copyright)

| Type of cement | Type of coarse aggregate | Compressive strengths ($N/mm^2$) | | | |
|---|---|---|---|---|---|
| | | Age (days) | | | |
| | | 3 | 7 | 28 | 91 |
| Ordinary Portland (OPC) or sulphate-resisting Portland (SRPC) | Uncrushed | 22 | 30 | 42 | 49 |
| | Crushed | 27 | 36 | 49 | 56 |
| Rapid-hardening Portland (RHPC) | Uncrushed | 29 | 37 | 48 | 54 |
| | Crushed | 34 | 43 | 55 | 61 |

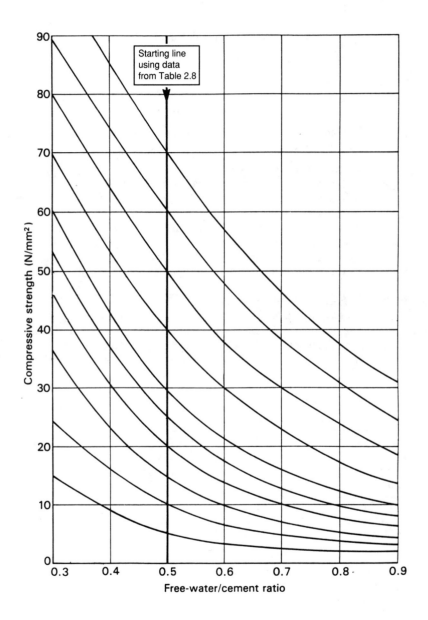

Fig. 2.13 Relationship between compressive strength and free-water/cement ratio. (Crown copyright.)

*Section 2. Workability*

    *2.1* Workability is specified by the slump or Vebe time of the fresh concrete.

    *2.2* From the value required and the aggregate shape and maximum size, the water content per cubic metre of concrete is obtained (Table 2.9).

*Section 3. Cement content*

    *3.1* It will be evident that, since free water/cement ratio and water content per cubic metre are now known and

$$\text{Cement content per cubic metre} = \frac{\text{Water content per cubic metre}}{\text{Free water/cement ratio}}$$

the cement content is easily calculated.

    *3.2* A maximum cement content may be specified (say to limit heat of hydration) in which case the cement content from 3.1 will need to be checked that it is not in excess of this value.

    *3.3* Similarly, a minimum cement content may be specified (say for durability ), so that the figure from 3.1 must be in excess of any such value. If it is not, the value specified must be used and the modified free water/cement ratio calculated.

*Section 4. Fresh concrete density*

    This is important as a part of the design process because it affects the yield of the concrete. The DoE method uses a graphical method for calculation of fresh density; an alternative means is given after this explanation of the DoE method.

Table 2.9   Approximate free-water contents ($kg/m^3$) required to give various levels of workability. (Crown copyright)

| Slump (mm)<br>Vebe (s) | | 0–10<br>>12 | 10–30<br>6–12 | 30–60<br>3–6 | 60–180<br>0–3 |
|---|---|---|---|---|---|
| Maximum size of aggregate (mm) | Type of aggregate | Water content ($kg/m^3$) | | | |
| 10 | Uncrushed | 150 | 180 | 205 | 225 |
|    | Crushed   | 180 | 205 | 230 | 250 |
| 20 | Uncrushed | 135 | 160 | 180 | 195 |
|    | Crushed   | 170 | 190 | 210 | 225 |
| 40 | Uncrushed | 115 | 140 | 160 | 175 |
|    | Crushed   | 155 | 175 | 190 | 205 |

*4.1, 4.2*  If the relative density of the aggregate (usually based on the saturated surface dry, SSD, condition) is known, the fresh density can be related to the water content of the mix, as in Fig. 2.14 .

*4.3*  The aggregate content per cubic metre of concrete is then:

Fresh density − (Cement content + Water content)

## Section 5. Proportion of sand

*5.1, 5.2*  The proportion of sand to total aggregate depends on the grading of the sand, as indicated by the proportion of sand passing the 600 µm sieve. It also depends on the maximum aggregate size, the workability and the free water/cement ratio. (Smaller maximum aggregate size, higher workability and higher water/cement ratio each increases the sand requirement.) The DoE method gives a proportion of sand corresponding to 'normal' cohesion. If higher or lower values are required, the percentage given could be increased or decreased by, say, 5 per cent respectively. Note that there are 12 sets of graphs in Fig. 2.15 − it is important to use the one appropriate to the aggregate size and workability chosen.

*5.3*  Sand content  =  Percentage of sand
                        ×  Total aggregate content

*5.4*  Finally, the coarse aggregate content is found by subtraction of sand from total aggregate.

---

## Worked example

Produce batch quantities for 1 cubic metre of concrete to have a slump of 30–60 mm and characteristic 28-day compressive strength of 25 N/mm² (5 per cent failures permitted). No previous results are available. The following materials are to be used:

| | |
|---|---|
| Cement | Ordinary Portland |
| Sand | Crushed, relative density 2.7; 55 per cent passes the 600 µm sieve |
| Coarse aggregate | Crushed, max. size 20 mm, relative density 2.7 |

This example has been worked on the standard design form in Fig. 2.16. Note:

1.  The figure obtained from Table 2.8 is 49 N/mm² corresponding to OPC, crushed aggregate and 28-day strength specification. This figure is used as the starting point on Fig. 2.13.

2.  Provision is made for aggregates to be divided into single sizes in the final summary of Fig. 2.16. This is only used for high-quality concrete and has not been carried out in the worked example. Where this is required the aggregates can be divided as follows:

    1 : 2      for 20 mm maximum size in 10 and 20 mm stockpiles;
    1 : 1.5 : 3  for 40 mm maximum size in 10, 20 and 40 mm stockpiles.

---

48

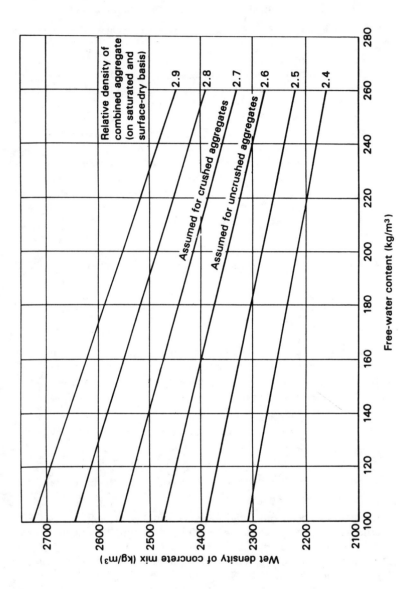

Fig. 2.14  Estimated wet density of fully compacted concrete. (Crown copyright.)

49

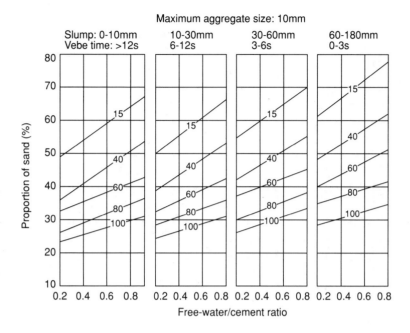

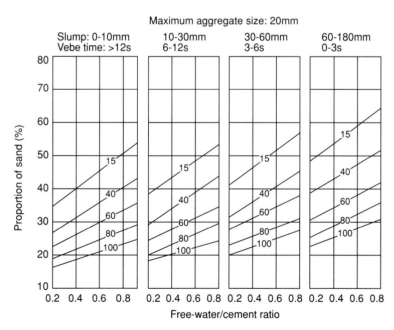

Fig. 2.15 Recommended proportions of sand according to percentage passing a 600 μm sieve. (Crown copyright.)

50

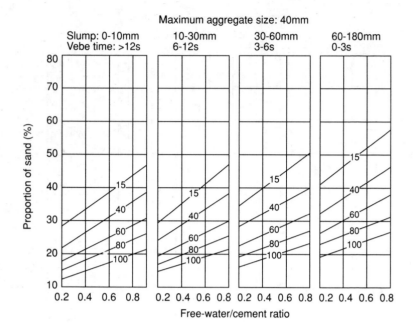

Fig. 2.15    Continued.

Concrete mix design form

Job title...... *Worked Example* ......

| Stage | Item | Reference or calculation | Values |
|---|---|---|---|
| 1 | 1.1 Characteristic strength | Specified | _25_ N/mm² at _28_ days |
| | | | Proportion defective _5_ % |
| | 1.2 Standard deviation | Fig 2.12 | _____ N/mm² or no data _8_ N/mm² |
| | 1.3 Margin | Table 2.7 | (k = _1.64_ ) _1.64_ × _8_ = _13_ N/mm² |
| | | | _____ N/mm² |
| | 1.4 Target mean strength | | _25_ + _13_ = _38_ N/mm² |
| | 1.5 Cement type | Specified | OPC/~~SRPC~~/~~RHPC~~ |
| | 1.6 Aggregate type: coarse | | Crushed/~~uncrushed~~ |
| | Aggregate type: sand | | Crushed/~~uncrushed~~ |
| | 1.7 Free-water/cement ratio | Table 2.8 Fig 2.13 | _0.58_ |
| | 1.8 *Maximum free-water/cement ratio* | *Specified* | _____ } Use the lower value  _0.58_ |
| 2 | 2.1 Slump or Vebe time | Specified | Slump _30–60_ mm or Vebe time _____ s |
| | 2.2 Maximum aggregate size | Specified | _20_ mm |
| | 2.3 Free-water content | Table 2.9 | _210_ kg/m³ |
| 3 | 3.1 Cement content | | _210_ ÷ _0.58_ = _362_ kg/m³ |
| | 3.2 *Maximum cement content* | *Specified* | _____ kg/m³ |
| | 3.3 *Minimum cement content* | *Specified* | _____ kg/m³ |
| | | | use 3.1 if ≤ 3.2 |
| | | | use 3.3 if > 3.1   _____ kg/m³ |
| | 3.4 Modified free-water/cement ratio | | _____ |
| 4 | 4.1 Relative density of aggregate (SSD) | | _2.7_ known/assumed |
| | 4.2 Concrete density | Fig 2.14 | _2400_ kg/m³ |
| | 4.3 Total aggregate content | | _2400_ – _210_ – _362_ = _1828_ kg/m³ |
| 5 | 5.1 Grading of sand | Percentage passing 600µm sieve | _55_ % |
| | 5.2 Proportion of sand | Fig 2.15 | _36_ % |
| | 5.3 Sand content | } | _0.36_ × _2400_ = _864_ kg/m³ |
| | 5.4 Coarse aggregate content | | _1828_ – _864_ = _964_ kg/m³ |

| Quantities | Cement (kg) | Water (kg or L) | Sand (kg) | Coarse aggregate (kg) | | |
|---|---|---|---|---|---|---|
| | | | | 10mm | 20mm | 40mm |
| per m³ (to nearest 5 kg) | 360 | 210 | 865 | — | 964 | — |
| per trial mix of _____ m³ | | | | | | |

Items in italics are optional limiting values that may be specified

OPC = ordinary Portland cement; SRPC = sulphate-resisting Portland cement; RHPC = rapid-hardening Portland cement. SSD = based on a saturated surface-dry basis.

Fig. 2.16   Concrete mix design form.

**Trial mixes**

It is most important to appreciate that no design method can produce exactly the right combination of fresh and hardened properties every time; a trial mix should always be made to ensure that the properties are as required with the particular materials and equipment used. The figures used in the design are based on 1 cubic metre of concrete, which would be excessively large for a trial mix. Hence the form of Fig. 2.11 provides for the calculation of what would normally be smaller batch quantities, for example, 50 litres for a trial mix. To obtain batch quantities for 50 litres ($0.05\,m^3$), the batch quantities for one cubic metre would be multiplied by 0.05.

Concrete which is produced by the trial mix would be tested for workability and fresh density and cubes would be made to check strength. If (a) workability or (b) strength are slightly in error, corrections can be made as follows:

(a) The correct water content for workability can often be judged by experience. If, for instance, in the worked example given, the actual water per cubic metre is found to be $220\,kg/m^3$, the cement content must be increased to keep the water/cement ratio and hence the strength constant. Hence the new cement content would be

$$\frac{220}{0.58} = 379\,kg/m^3$$

The fresh density (if not measured) would also be expected to reduce slightly. Figure 2.14 indicates that the correct value would be approximately $2390\,kg/m^3$. The new total aggregate content is therefore

$$2390 - (220 + 379) = 1791\,kg/m^3$$

The sand content would be found as previously.

(b) The point on Fig. 2.13 corresponding to the strength *obtained* and the water/cement ratio used should be located. Then proceed parallel to the curved lines until the desired strength is obtained and read the new water/cement ratio. If, for instance, in the worked example, a strength of $35\,N/mm^2$ was obtained instead of $38\,N/mm^2$, find the point corresponding to $35\,N/mm^2$ and 0.58 water/cement ratio. Moving upwards parallel to the curved lines until $38\,N/mm^2$ is reached and then reading vertically produces a water/cement ratio of 0.55. The remaining mix details are now recalculated.

Where both workability and strength are incorrect, then both a new water content and new water/cement ratio would be obtained and then the method completed as before.

Second trial mixes should only be necessary where substantial errors are involved, though a final full-scale trial mix is advisable before going into production. Adjustments for moisture in aggregates are, of course, essential if trial mixes are to fulfil their function.

## Calculation of concrete fresh density

It is only from an accurate knowledge of concrete density that batch masses of cement and aggregates for a given volume of concrete can be reliably predicted. The following technique is therefore given and serves as an alternative to the method used in the DoE method of mix design.

Suppose, for example, $400\,m^3$ of concrete is to be produced to the specification given on page 51. The batch quantities per cubic metre are:

| Water | 210 litres | Sand | 865 kg |
| Cement | 360 kg | Coarse aggregate | 964 kg |

The relative density of cement is required and this is assumed to be 3.15.

The figures are tabulated as in Table 2.10 and, to simplify them, all figures are divided by the mass of cement, giving masses which would correspond to 1 kg of cement.

The volumes of each component in litres are easily obtained by dividing the reduced *mass* of the component by its *relative density*. Then

$$\text{Concrete density} \quad = \quad \frac{\text{Total mass}}{\text{Total volume}}$$

$$= \quad \frac{6.664}{2.782}\,\text{kg/litre}$$

$$= \quad 2.395\ \text{kg/litre}$$

$$\text{or, to nearest } 10\,\text{kg/m}^3 \quad = \quad 2400\ \text{kg/m}^3$$

(This figure agrees with the value obtained from Fig. 2.14.) To obtain the batch quantities for $400\,m^3$ of concrete, the batch quantities per $m^3$ are simply multiplied by 400. If, as may be the case, mass ratios only are given, the ratios are multiplied by the appropriate factor (2400/6.664 in this case) to give batch quantites per cubic metre.

The procedure is then as before.

Table 2.10   Illustration of the method used for calculating fresh density

| Material | Mass (kg) | Mass ratio (per kg cement) | Rel. density (SSD) | Volume (litres) |
|---|---|---|---|---|
| Water | 210 | 0.583 | 1.00 | 0.583 |
| Cement | 360 | 1 | 3.15 | 0.317 |
| Sand | 865 | 2.403 | 2.70 | 0.890 |
| Coarse aggregate | 964 | 2.678 | 2.70 | 0.992 |
| Total | 2399 | 6.664 | | 2.782 |

If the presence of air is to be allowed for in calculating density, this is easily done by increasing the total volume of concrete, but not the mass, by the percentage of air that is expected.

## Statistical interpretation of cube results

In concluding the section on the design of concrete, we return to the subject of variability. The reasons for strength variations have already been given and it will be evident that the extent of variations will depend to a large degree on the quality control procedures carried out on any one site. Figure 2.17 shows, for example, two histograms relating to 50 cube tests on two sites A and B. It will be evident that curve A, although of lower mean strength, is much narrower than curve B and that, with site A, unlike site B, there are no cube results below $30 \, \text{N/mm}^2$. The curves have been drawn such that characteristic strengths based on 8 per cent failures are the same, though site A satisfies this requirement with a lower mean strength (hence a more economical concrete) and fewer very low cube results. The standard deviation ($s$) of a set of $n$ results, values $x_1, x_2, \ldots, x_i, \ldots, x_n$ of average value $\bar{x}$ is given by

$$s = \sqrt{\left[ \frac{\Sigma(x_i - \bar{x})^2}{n-1} \right]}$$

or, to put it alternatively,

$$s = \sqrt{\left[ \frac{\text{Sum of squares of (each individual result minus mean strength)}}{n-1} \right]}$$

The value of $s$ obtained in practice for ordinary concrete of mean strength $40 \, \text{N/mm}^2$ might vary from $4 \, \text{N/mm}^2$, which represents good quality control, to $8 \, \text{N/mm}^2$, which represents poor quality control.

Perhaps the most direct way of checking that the characteristic strength is satisfactory is to keep calculating the mean strength ($f_m$) and standard deviation ($s$) using, say, the most recent 50 results and then to calculate characteristic strength ($f_c$) by the equation

$$f_c = f_m - ks$$

$k$ being the factor appropriate to the permissible number of failures, as given in Table 2.7. Such a method would, however, be mathematically quite arduous and would indicate rather slowly any change in quality control. It is especially important that any adverse change in strength be readily detectable as soon as possible and one convenient technique is the use of simple quality control charts in which strengths of consecutive cubes are represented as points. It is often found more convenient in practice to take means of four cubes and to plot the means, since this reduces the variability of results about the mean strength.

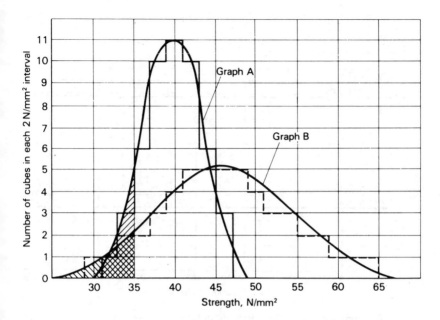

Fig. 2.17 Histograms drawn from the results of two sets of 50 concrete cube tests. Graph A represents good-quality control and graph B poor-quality control. The total area under each graph is the same. Also, the area under each curve to the left of the 35 N/mm² line is the same, implying equal numbers of results beneath this strength. Curve B is, however, indicative of a greater probability of very low cube results than curve A.

Suppose, for example, the required characteristic strength is 30 N/mm² with 5 per cent failures allowed and standard deviation equal to 6 N/mm². The target mean strength would then be

$$\begin{aligned} f_m &= f_c + ks \\ &= 30 + 1.64 \times 6 \\ &= 39.84 \text{ (say 40) N/mm}^2 \end{aligned}$$

If means of 4 are plotted, the standard deviation of the *means* should be

$$\frac{\text{Standard deviation of individual results}}{\sqrt{[\text{Number in group}]}} = \frac{6}{\sqrt{4}} = 3 \text{ N/mm}^2$$

Hence not more than 5 per cent of means of 4 should be below

$$40 - 1.64 \times 3 = 35.08 \text{ (say 35) N/mm}^2$$

A chart could then be constructed, as in Fig. 2.18, with mean strength

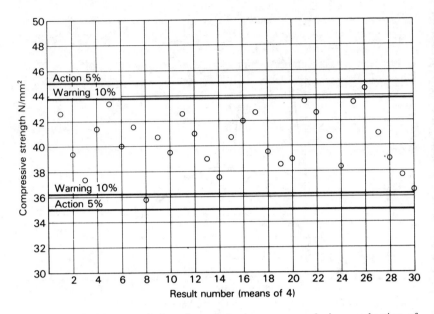

Fig 2.18  Simple control chart for monitoring progress during production of concrete. In this case progress is satisfactory since there are no results outside 'Action' limits and results are equally distributed about the mean.

$40 \text{ N/mm}^2$ and 'action lines' at 35 and $45 \text{ N/mm}^2$. If there is a definite trend towards, say, the $35 \text{ N/mm}^2$ line, then action, in the form of increasing the mean strength of the concrete, must be taken, since the points should be evenly spaced about the means with no more than one result in 20 (5 per cent) below the lower line if concreting is proceeding satisfactorily. 'Warning' lines may also be included corresponding to, say, 10 per cent limits. Since, for 10 per cent failures, $k$ is 1.28 (Table 2.7), the lines would be at strengths of

$$40 \pm 1.28 \times 3 \ = 43.8 \text{ and } 36.2 \text{ N/mm}^2 \text{ approximately.}$$

The warning lines would serve to indicate that the concrete quality is varying. Increases in mean strength should also be checked, since unnecessarily high mean strengths (or reductions in standard deviation) may allow a reduction in cement content of the concrete, hence saving cost. Small corrections to strength are, in fact, most commonly made by small alterations to cement content; a simple guide is that, to increase mean strength by $0.8 \text{ N/mm}^2$, the cement content should be increased by $5 \text{ kg/m}^3$ concrete and vice versa. The aim, in all concrete testing, is to obtain results as soon as possible – accelerated curing may be used to predict 28-day strength as early as one day after placing provided reliable correlations have been produced.

Hence, faulty concrete can be identified and removed before it is fully mature or further lifts have been cast. More elaborate charts may be used alternatively; for instance, a second chart plotting variability within groups of 4 cube results, compared with a 'target mean variability' might be used. This would reveal any change in quality control (represented by standard deviation) as work proceeds.

## Admixtures

Admixtures (BS 5075) are materials other than cement, water or aggregates added at the mixer. Their use has increased greatly in recent years but, although there are many instances in which marked improvements in properties for specific applications are obtainable, careful precautions regarding their use are advised for the following reasons:

- Doses are usually very small – often less than 1 per cent by weight of cement, hence, careful batching is necessary, the use of special dispensers being recommended.
- Effects are sometimes very sensitive to variations in dose, hence good supervision is essential.
- Side effects are often present, especially if overdoses occur.
- To ensure effective dispersion, liquid admixtures should be added to the mixing water rather than to the mixer and water-soluble solid admixtures should be dissolved in the mixing water before use.

There are four main categories of admixture, though in practice it may be convenient to combine several effects within one admixture; in particular, most admixtures are formulated to have a plasticising action since there is very little increase in cost needed to achieve this. Most admixtures are now manufactured to BS 5075 and with careful use can be very beneficial to concrete for specific applications.

### Water reducers (plasticisers)

These are perhaps the most widely used admixtures in the concrete industry, their use in ready-mixed concrete being especially common.

The main water-reducing compounds are lignosulphonates and hydrocarboxylic acids, and they operate by attaching themselves to cement grains and imparting a negative charge, which causes grains to disperse more effectively. Hence the formation of 'flocs' (groups of cement particles), which tend to trap mixing water, is avoided (Fig. 2.19). Water-reducing admixtures may be used to:

(a) allow a water reduction, hence reducing water/cement ratio and increasing strength; or

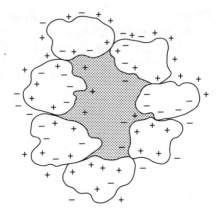

Fig. 2.19   Cement particles, on account of surface charges, form into 'flocs', trapping water and reducing workability.

(b)  produce an increase in workability at given water and cement content; or
(c)  allow a reduction in water and cement contents in concrete of a given strength, hence reducing cost.

Typical water reductions are in the region of 5–15 per cent when used as in (a) or (c) above. Side effects include air entrainment and retardation of set, though these are only slight at normal doses. Plasticisers have relatively low cost and form an economical means of improving concrete quality if used as in (a) above. However when used as in (c) above, the reduced cement content could lead to durability problems.

More recently, a number of 'super-plasticisers', based on compounds such as sulphonated melamine formaldehyde, have been introduced. These can be used at higher dosages than ordinary water reducers without the same side effects. Since they are quite expensive, they are used mainly for applications (a) and (b) above. In the latter, the concrete, which has a slump of over 150 mm, is often referred to as 'flowing concrete' since it is more or less self-levelling and self-compacting, reducing labour costs in applications such as large floor slabs. The effect of super-plasticisers is short lived (only 30 to 45 minutes), hence they are usually added to the concrete just before placing. Care is necessary to avoid segregation but hardened properties, such as strength, shrinkage and creep, do not appear to be adversely affected.

## Retarders

The object of these would normally be to retard the set (rather than the strength development) of the concrete. Hence their effect should last a few hours only.

They are normally based on lignosulphonates, hydrocarboxylic acids or sugars. They act mainly on those compounds in cement which are responsible for setting and early strength; that is the $C_3A$ and $C_3S$. Uses include the following:

• To reduce the setting rate when concreting in hot weather. In many cases the retarder can improve long-term strength when used in this way, since slower hydration of cement leads to an improved 'gel' structure.

• In very large pours, as, for example in mass concrete foundations in order to produce monolithic (structurally continuous) concrete in spite of a prolonged placing period.

• An overdose may result in a greatly increased setting time, though the concrete will eventually set and there should be no long-term effect. A large overdose (which may still only be 0.5 per cent by weight of cement) will 'kill' the set; retarders can be used, for example, to avoid hardening of concrete in the drum of a ready-mix agitator truck which has broken down.

## Accelerators

These are designed to accelerate the early strength development of concrete. They are traditionally based on calcium chloride, which increases the hydration rate of the calcium silicates and, to some extent, the $C_3A$ in the cement. Substantial improvements can be obtained between the ages of 1 day and 7 days, especially in cold weather – for example, the 3-day strength of concrete may be doubled at a temperature of 2°C. Long-term strength is unaffected by the use of calcium chloride. Most important are the side effects of the material – the susceptibility of steel in structural concrete to corrosion is greatly increased in the presence of chlorides, such that BS 8110 now limits chlorides, measured as chloride ion content, to 0.1 per cent by weight of cement in prestressed concrete or steam-cured concrete, or 0.2 per cent when sulphate-resisting cement is used. The maximum permitted chloride ion content in other steel-containing concretes is 0.4 per cent. (These chloride ion contents are equivalent to approximately 50 per cent greater percentages of anhydrous calcium chloride by weight.) Any chorides present in the aggregates – for example, from sea-dredged sources – must be included in the above figures. The shrinkage of concrete is also increased by the presence of chlorides. Chloride-free accelerators are now available but these are more expensive than choride-based accelerators and it should be noted that more rapid strength development can be obtained by other means; for example, use of water-reducing agents or increased cement content.

## Air-entraining agents

These are based on substances such as vinsol resins, which reduce the surface tension of the water, allowing stable air bubbles to be introduced by mixing. These air bubbles are much smaller than *pockets* of entrapped air; most are less than 1 mm in diameter and they do not admit water, even in saturated conditions. The uses of air-entraining agents are based on the following properties of air-entrained concrete:

1. Resistance of hardened concrete to frost and de-icing salts is greatly improved. The close distribution of small air bubbles throughout the mortar fraction of the concrete appears to act as a means of pressure relief as freezing occurs in the concrete. Current specifications for concrete roads in the UK stipulate an air content of 4.5 per cent in the top 50 mm of the concrete.

2. Resistance to segregation and bleeding is improved, together with workability, hence air-entrained concrete could be used to improve harsh mixes or to permit placing in conditions where segregation might be a problem.

The air present reduces the concrete strength, though at least part of this shortfall can be avoided by use of lower water contents for a given workability.

## Curing of concrete

Curing is the process by which cement is enabled to hydrate. Since water is involved in the process it is important to maintain the concrete in a saturated condition for a sufficient length of time to develop a satisfactory hydrate structure. The minimum period for this is usually considered to be about one week, though it may be longer with cements which hydrate slowly such as pozzolanic types. The temperature is also an important factor, the 'ideal' value being between 5 and 15 °C since hydration is very slow at lower temperatures, while high temperatures tend to cause evaporation, together with an inferior 'gel' structure. However, with care, concrete can be cured at temperatures up to about 30 °C. In no circumstances must concrete be allowed to freeze before it reaches maturity; the accompanying expansion of the water could cause serious surface damage.

As an aid to memory it may be added that to produce good concrete the four 'C's should be considered:

Cement content
Cover to steel reinforcement
Compaction
Curing

## Lightweight concretes

These may offer the following advantages over ordinary dense concrete:

- Since less raw material is involved, it may be argued that lightweight concretes help conserve materials resources.
- A number of lightweight aggregates are produced from what are otherwise waste materials – for example, pulverised fuel ash and blast-furnace slag.
- Foundation loads are reduced, hence foundation size can be decreased.
- Higher lifts can be cast for a given formwork system, since formwork pressures are lower. Hence, the number of construction joints can be reduced.
- Some lightweight concretes, such as precast aerated blocks and no-fines concrete, provide an excellent key for rendering and plastering.
- Fixings are often more easily made than with dense concretes. Some types can be nailed (using cut nails).
- Thermal insulation is improved compared with dense aggregate concrete.
- Fire resistance of lightweight concretes is superior to that of most dense concretes.
- High-frequency sound absorption is better than that of dense concrete.
- Lightweight precast blocks are much easier to handle and cut than dense concrete equivalents.

There are three main methods of producing lightweight concretes, each method producing concrete with characteristic properties and applications.

## Lightweight aggregate concrete

A large range of concrete densities can be obtained using lightweight aggregates – from 500 to 2000 kg/m$^3$. The thermal conductivity of concrete decreases as density decreases, values ranging from about 1.0 W/m°C for the higher densities to under 0.2 W/m°C for concrete of very low density. Strengths decrease similarly with density, though certain aggregates may give higher strength than others at a given density. An aggregate would normally be chosen according to the density and strength required, subject also, of course, to local availability. Lower density concretes would be used in situations where, perhaps, thermal insulation properties rather than strength are important – for example, as insulating layers in floors or roofs. Concretes having densities of over 1000 kg/m$^3$ can, however, be used 'structurally' and there are now many structures incorporating reinforced and prestressed lightweight aggregate concrete. By choice of suitable aggregates, such as foamed slag or sintered pulverised fuel ash, concrete strengths of over 40 N/mm$^2$ are obtainable with densities in the range 1500–2000 kg/m$^3$. The substitution of natural sand for lightweight fine aggregate can be used to increase strength still further, though density will clearly rise as well. In designing lightweight concrete structures, allowance must be made for the lower elastic modulus and higher shrinkage and creep of lightweight aggregate concretes.

The production of lightweight aggregate concretes requires care because:

- Design is best tackled from volume considerations in conjunction with trial mixes. Hence, bulk densities must be known if, as is normally the case, weight batching is used.
- Aggregates often absorb a great deal of moisture – 15 per cent or more by weight. Hence, the effective free water/cement ratio, an important factor in controlling strength, is often difficult to predict.

## Aerated concrete

Aerated concrete comprises a cement/sand mortar into which gas is introduced. This may be obtained chemically, for example, by the use of aluminium powder, which produces hydrogen in the presence of alkalis released by the cement, or by the use of foaming agents. Careful control of plastic properties is essential in order that the gas aerates the mortar without escaping, hence, these concretes are mainly produced in the form of manufactured precast blocks. Densities in the range 400–800 kg/m$^3$ are obtained, corresponding to compressive strengths of 2–5 N/mm$^2$

Conventionally cured aerated concretes would have high shrinkage but this can be reduced by high-pressure steam curing (autoclaving). Hence many commercial precast blocks are autoclaved and these would also normally contain a pozzolanic material, such as pulverised fuel ash, which contributes to strength.

With increased standards of thermal insulation now being mandatory in residential buildings, the use of aerated concretes, which have very low thermal conductivity (in the region of 0.2 W/m°C) has increased greatly. Products are available for walling of low- to medium-strength requirement. Aerated blocks are reasonably frost-resistant, though they would normally be protected from the weather, if used externally, by cladding or rendering.

## No-fines concrete

As the name implies, this concrete is produced from cement and coarse aggregate only, the latter usually being graded between 20 and 10 mm. The aggregate may be ordinary dense aggregate, which would result in the concrete having a density in the region of 1800 kg/m$^3$, or lightweight aggregate, resulting in a concrete density as low as 500 kg/m$^3$. Dense aggregates would produce strengths of up to 15 N/mm$^2$, according to cement content, while the ceiling strength for lightweight aggregates would be around 10 N/mm$^2$.

In producing no-fines concrete, great care should be taken to obtain the correct water content, since a low value would result in uneven distribution of cement, while a high value would cause settlement of the cement paste fraction to the base of the concrete. Only minimal vibration should be used for compaction.

Shrinkage of no-fines concrete is low, since the cement paste is discontinuous, though shrinkage occurs quickly due to the high air permeability of the material. No-fines concrete is resistant to frost, though rendering of exposed surfaces is normally carried out to prevent water penetration.

No-fines concrete was formerly used as an *in-situ* walling material for low-rise dwellings. It is also manufactured in the form of lightweight precast blocks.

## High-strength concrete

The term 'high-strength concrete' generally refers to concretes having 28-day compressive strength over 60 N/mm$^2$. Such concretes may offer the following advantages:

- Increased hardening rates leading to reduced construction times.
- Smaller section sizes for a given load, leading to economies of space.
- Greater stiffness, hardness, chemical resistance and durability.

The DoE method of mix design can be used to produce high-strength concrete by use of one or more of the following modifications:

- Use super-plasticisers to obtain water/cement ratios of 0.3 or less.
- Produce rich mixes of very low workability. It should be noted that such mixes will require pressure/vibration for compaction and therefore have limited site application. Their use would also be confined to situations where increased shrinkage is not a problem – for example, in precast units.
- Use vacuum dewatering followed by power trowelling (for floors).
- High-strength precast units can be made by autoclaving mixes containing pozzolanas.

To obtain concrete of strength up to 100 N/mm$^2$ the use of high-strength aggregate such as crushed granite may be essential since low-strength aggregates or those having smooth surfaces will tend to initiate failure under stress. Such concretes are readily obtainable either site-mixed or ready-mixed. Production of strengths of over 100 N/mm$^2$ is likely to be confined to precasting works which have suitable facilities.

## Shrinkage and moisture movement of concrete

Shrinkage, or to use a fuller description, *drying* shrinkage, is a contraction occurring when concrete is dried for the first time. This should be distinguished from subsequent movements resulting from moisture changes and referred to as 'moisture movement' because the initial shrinkage is partly irreversible and is much more likely to result in damage to the concrete than subsequent movements. Figure 2.20 gives a schematic representation of typical volume changes of concrete with time. During wet curing there is a slight expansion – indeed, it should be noted that cement expands on hydration by over 100 per cent, filling at least part of the space previously occupied by water. Shrinkage occurs when, on drying, water from within the cement 'gel' is removed, causing gel surfaces to approach slightly. Hence, the origin of shrinkage lies within the hydrated cement and it follows that rich, wet mixes, which contain more cement gel, shrink more. Long-term shrinkage strains vary from $200 \times 10^{-6}$ for lean, low-workability mixes, to over $700 \times 10^{-6}$ for rich, wet concretes. Approximately 40 per cent of this movement is irreversible – the concrete does not regain this proportion if rewetted. Moisture movement does, however, occur on each subsequent wetting and drying, as water is admitted to and removed from the cement gel. Even the lower shrinkage value given above could cause tensile failure in the concrete if shrinkage is restrained or if differential shrinkage occurs between different parts of a structure since the tensile strain capacity of mature concrete is only about $100 \times 10^{-6}$ and that of young concrete is much lower still. Hence, it is important to cure concrete until it has some ability to withstand the strains occurring due to shrinkage. Examples of situations in which shrinkage must in any case be allowed for are:

- Ground floor slabs, hard standings and roads, where subgrade friction can generate stresses sufficient to cause failure unless joints are provided.
- Mass concrete, where surface shrinkage is restrained by inner layers. Surface shrinkage reinforcement can be provided to control this.
- Clay brickwork infill panels in reinforced concrete frames – especially if the bricks are fresh from the kiln, leading to a risk of expansion. Movement provision is essential.

The concrete frames of reinforced concrete buildings do not normally suffer greatly from shrinkage since the contraction is largely unrestrained.


## Durability of concrete


In the 1980s, there was increasing concern over the rate of deterioration of some concrete structures. This, together with the fact that concrete buildings can be very expensive to repair, and a revitalised steel industry, led to a decline in the use of concrete for some applications. In most instances, deterioration of concrete could be linked to failure to implement existing technology; for example, due to lack of supervision, though in a few instances, decay has been the result of inadequate understanding of the behaviour of the material. There is now a greater awareness of potential problems and increased confidence in the performance of concrete generally.

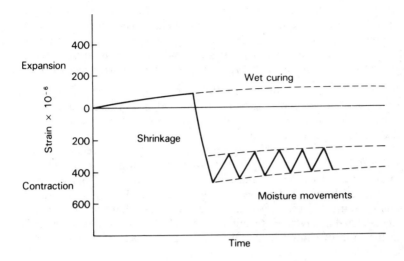

Fig. 2.20   Shrinkage and moisture movements in concrete.

Almost all forms of deterioration are the result of water ingress; hence, the need to keep water out of structures, or to take special care where this is not possible – and this cannot be overstated. The action of water may be twofold:

1.  Physical attack, where water crystallises on freezing or where drying out causes damage due to crystallisation of dissolved salts in the water.
2.  Chemical damage. Most forms of chemical damage are associated with water since water is essential for the formation of ions – the chemically active form of salts or other materials.

The main types of deterioration are given below.

### Frost attack

There are three prerequisites for frost attack:

1.  A permeable form of concrete able to admit water.
2.  The presence of water (though materials need not be saturated to be affected by frost).
3.  Temperatures below 0 °C.

The mechanism of attack has long been attributed to the 10 per cent expansion of absorbed water on freezing, though it is now known that further damage can result from movement of water within concrete on cooling below 0 °C, ice building up in large pores and cracks and causing very large expansion in local areas while other areas become dryer (desiccation). Perhaps the extreme example of this is frost heave, as found in soils where large ice lenses can be obtained. Frost damage in concrete is exacerbated by:

* Horizontal surfaces which not only tend to absorb more water in wet conditions but also cool quicker by radiation to the sky on a clear night.
* Very low temperatures which increase the extent of migration of water and result in freezing to greater depths in the concrete.
* Repeated freezing and thawing rather than continuous freezing. (Once freezing has occurred no further damage is done at a steady low temperature. However, a thaw followed by a further frost will start another cycle of damage. Hence climates as found in the northern UK could be more damaging than colder climates where thawing is less frequent and in any case precipitation in such climates is more likely to be in the form of snow which, being solid, cannot penetrate and therefore cannot damage concrete, at least until thawing occurs.)
* Weak, permeable concretes which absorb water more readily.
* Use of de-icing salts which melt the ice, adding to saturation of the concrete and then crystallise adding to the damage.

Most of the measures required to avoid frost attack will be evident from the above, though it should be added that the use of air-entraining agents has been found to be particularly effective in avoiding damage to surfaces at risk. It is important to appreciate that new concrete with its incomplete hydrate structure and high permeability is particularly at risk from frost, and must be protected by an insulating material until it has sufficient maturity to resist frost.

## Sulphate attack

Sulphates are often found in the ground, and are mainly associated with clay soils. To be active they must be in solution so that the risk of attack depends not only on the salt content of a soil but on the presence and movement of moisture. The principal mechanism is one in which sulphate ions react with hydrated $C_3A$ in the cement, producing an expansive product, calcium sulphoaluminate, which disrupts the concrete. Sulphates commonly occur in three forms – calcium, sodium and magnesium, which pose quite different risks. Calcium has low solubility (about 1.4 mg of ions per litre) so that it does not constitute a high risk. Of the other two, magnesium is more harmful because:

* The reaction product is insoluble, precipitating out of solution and leaving the way clear for continued attack.
* Magnesium sulphate also reacts with the $C_3S$ hydrate in cement.

When estimating the risk of sulphates it is therefore important to identify the type of sulphates present. Total sulphate content could be misleading if it were mainly in the form of calcium sulphate. Careful sampling is also necessary since levels may vary widely from place to place and ground water may be diluted during sampling or after wet weather. The best procedure is normally analysis of ground water or water obtained from a 2:1 water/soil mixture.

For ground water containing up to 1.4 g/litre of $SO_4$ ions, OPC can be used provided a cement content of at least 330 kg/m$^3$ is used with water/ cement ratio not higher than 0.5. At sulphate levels up to 6.0 mg/litre without high magnesium concentrations, use of pozzolanas at high replacement levels, or sulphate-resisting Portland cement, would be necessary. At higher sulphate levels, or when high magnesium levels are found, the use of sulphate-resisting Portland cement in conjunction with a waterproof coating would be essential. In all cases it is of paramount importance that attention is given to achieving full compaction.

## Crystallisation damage

This may occur in concretes which are resistant to chemical attack. Sulphates (or other salts) may be admitted by permeable concretes and if, at a later stage, drying occurs, these salts crystallise, causing damage similar to that caused by frost. The condition may be severe when concrete is subject to wetting and drying cycles – as, for example, in sea walls around the high tide mark. Attack is also serious if one area of a structure is saturated while an adjacent area is dry – salts migrate towards the dry area, causing extensive crystallisation in the region of drying. This type of damage may also be avoided by the use of low water/cement ratio concretes, fully compacted to produce low permeability.

## Alkali silica reaction (ASR)

Although relatively uncommon at the moment ASR, sometimes called 'concrete cancer', has evoked a strong public response because it is a relatively new form of deterioration and there is no effective repair technique for structures, once seriously affected. ASR occurs when three conditions are met:

1.  The concrete has a very high pH value (over 12) resulting from the presence, in quite small quantities, of sodium and potassium alkalis originating from the oxides of these elements if present in the cement (high-alkali cements).
2.  Aggregates containing reactive forms of silica such as found in opal, tridymite, some flints and cherts.
3.  Water is present.

The reactive components of the aggregates form a gel leading to expansion and disruption of the concrete, often producing 'map' cracking (Fig. 2.21).

Fig 2.21 'Map cracking' in concrete affected by ASR. Cracks tend to meet in groups of three. In this photograph the cracks are highlighted by absorbed moisture.

The reaction is slow, and cracking usually takes 10 to 20 years to become evident. There is a 'pessimum' level of about 2 per cent of reactive aggregate since smaller quantities do not form sufficient gel while larger quantities 'swamp' the excess alkali present. Once formed, cracks admit further water, accelerating the rate of deterioration. ASR can be avoided by the following:

- *Use of low-alkali cements.* For example, low-alkali versions of OPC (BS 12) or SRPC (BS 4027) containing less than 0.6 per cent alkali expressed as $Na_2O$ equivalent. Alternatively, pozzolanic additions are known to be beneficial in reducing ASR.
- *Avoidance of reactive aggregates.* Some types, such as granite or limestone, are unlikely to be reactive. In areas of doubt, expansion tests can be carried out on samples of concrete made from the aggregates in question placed in alkaline solutions at 38 °C for some weeks.
- *Exclusion of water.* Most structures affected to date have been severely exposed, for example, concrete bridges or buildings near the sea. The risk to interior concrete is very small except in swimming pools or other damp areas. Once attack has commenced, it may be very difficult to exclude further water from affected structures.

## Corrosion of steel reinforcement

This is one of the commonest problems in structural concrete. Where damage is localised, repair is possible, though in view of the labour-intensive nature of the repair process, it can be very expensive. Extensive corrosion of reinforcement may result in demolition. The corrosion of steel in moist environments is described elsewhere (p. 00). In concrete, steel will be protected, provided an alkaline environment exists, since a protective layer of $Fe_2O_3$ (ferric oxide) is formed and there is no tendency to ionise (corrode). Such an environment is provided by calcium hydroxide, one of the products of hydration. This material is, however, attacked from the surface by carbon dioxide in normal moisture-containing atmospheres, producing almost neutral calcium carbonate:

$$Ca(OH)_2 + CO_2 \rightleftharpoons CaCO_3 + H_2O$$

The rate of attack and hence the depth of penetration of this 'carbonation' reaction depend chiefly on the permeability of the concrete. Well compacted, low water/cement ratio concretes have low permeability such that carbonation would only reach, perhaps, 5 mm depth – much less than the depth of normal steel reinforcement – in many years. In lower quality concretes, much higher carbonation depths of 25 mm or more may occur, putting steel at risk. The need to fix steel accurately in position at the time of construction in order to give consistently sufficient cover will be apparent.

The risk of corrosion to steel in existing structures can be easily checked by breaking off pieces of concrete from the surface and spraying the freshly exposed surface with a phenolphthalein solution. This indicator turns pink in alkaline areas revealing clearly the carbonation depth (Fig. 2.22). Depths of reinforcing bars can be ascertained by using a covermeter, based on the magnetic effect of steel. Unless the steel is well beneath the carbonated layer, there is a risk that corrosion could occur.

Where chlorides are present; for example, from admixtures used in the concrete, or de-icing salts, the corrosion risk is greatly increased since chloride ions have a de-passivating effect and also increase the electrical conductivity of the concrete. Their use is now strictly controlled in structural concrete (see page 59).

The risk of reinforcement corrosion can be reduced by:

• Utilising anti-carbonation coatings (usually also decorative) on the concrete surface to reduce the rate of carbon dioxide ingress.
• Using special steels for reinforcement; for example, in order of increasing cost, galvanised, epoxy coated and stainless steel.

The best means of achieving longevity, nevertheless, remains the use of good-quality concrete with careful supervision to ensure that cover to steel is always adequate.

Fig. 2.22   Phenolphthalein test on precast concrete post used in prefabricated housing. The alkaline areas are darker.

## Experiments

## Experiment 2.1   Grading of aggregates (BS 812: Section 103.1, BS 882)

Note that the term 'sand' rather than 'fine aggregate' is used in the experiment to signify material passing a 5 mm sieve in accordance with BS 882 (1992).

*Apparatus*
BS sieves; weighing balance accurate to 1 g.

*Sample preparation*
A main sample is first obtained from the stockpile by the procedure outlined in BS 812, Part 101. The essential requirement is that this sample should consist of at least 10 increments, obtained from different parts of the stockpile and mixed thoroughly. The main sample should be of not less than 13 kg for sand and not less than 25 kg for aggregates of nominal size between 5 and 28 mm.

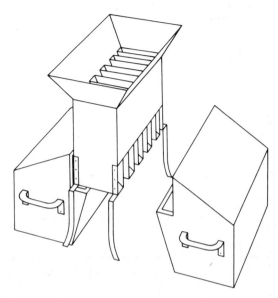

Fig. 2.23   Riffle box for reduction of aggregate sample size. Three receiving boxes are normally supplied.

The main sample is reduced in size to a quantity suitable for sieving, using a riffle box (Fig. 2.23), or by quartering. It is recommended that division be continued until the minimum mass in excess of the requirements of Table 2.11 is obtained; larger masses, paticularly of sand, tend to lead to blinding of sieve apertures. Note, however, that shaking a divided sample from a scoop to obtain the exact minimum mass of Table 2.11 is incorrect as this leads to particle segregation. The prepared sample must then be dried either in an oven or by being placed for some time on a flat sheet.

*Procedure*
Select the sieves appropriate to the nominal size of material being used by reference to Tables 2.12 and 2.13. Examples are:

| | |
|---|---|
| 20 mm graded coarse aggregate: | 37.5, 20, 14, 10 and 5 mm |
| 20 mm single-size coarse aggregate: | 37.5, 20, 14, 10 and 5 mm |
| 10 mm single-size coarse aggregate: | 14, 10, 5 and 2.36 mm |
| sand: | 10, 5, 2.36 and 1.18 mm |
| | 600, 300 and 150 mm |

Shake the material through each sieve individually into a collecting tray starting with the largest, using a varied motion, backwards and forwards, clockwise and anticlockwise and with frequent jarring. (Alternatively, sieves

Table 2.11   Minimum mass of sample for sieve analysis

| Nominal size of material (mm) | Minimum mass of sample (kg) |
|---|---|
| 50 | 35 |
| 40 | 15 |
| 28 | 5 |
| 20 | 2 |
| 14 | 1 |
| 10 | 0.5 |
| 6 | 0.2 |
| 5 | 0.2 |
| 3 | 0.2 |
| <3 | 0.1 |

Table 2.12   Gradings for sand

| BS 410 test sieve | Percentage passing BS sieves | | |
|---|---|---|---|
| | F | M | C |
| 10 mm | 100 | 100 | 100 |
| 5 | 89–100 | 89–100 | 89–100 |
| 2.36 | 80–100 | 65–100 | 60–100 |
| 1.18 | 70–100 | 45–100 | 30–90 |
| 600 µm | 55–100 | 25–80 | 15–54 |
| 300 | 5–70 | 5–48 | 5–40 |
| 150 | 0–15 | 0–15 | 0–15 |

Note that the figure of 15 per cent for the 150 µm sieve is increased to 20 per cent for crushed rock sands except when they are used for heavy duty floors.

may be assembled into a sieve shaker). In order to prevent blinding of sieve apertures, the quantity of material retained on sieves should not exceed the values given in Table 2.14. Smaller sieves are often a problem in this respect and one way of solving this is to use additional sieves – for example, a 425 mm sieve prior to the 300 mm sieve. The quantities retained on the sieves are then *added*. Weigh quantities retained on each sieve and check that the total is approximately equal to the original sample size. If it is not, repeat the procedure with a new sample.

Complete a table similar to that of Table 2.4 using appropriate sieve sizes. Plot the percentage passing each sieve using percentage passing as the y axis and sieve sizes increasing equally spaced as the x axis.

Table 2.13    Grading limits for coarse aggregates

| BS sieve (mm) | Percentage by weight passing BS sieves | | | | | | | |
| | Nominal size of graded aggregate (mm) | | | Nominal size of single-size aggregates (mm) | | | | |
| | 40–45 | 20–25 | 14–5 | 40 | 20 | 14 | 10 | 5 |
|---|---|---|---|---|---|---|---|---|
| 50.0 | 100 | – | – | 100 | – | – | – | – |
| 37.5 | 95–100 | 100 | – | 85–100 | 100 | – | – | – |
| 20.0 | 35–70 | 90–100 | 100 | 0–25 | 85–100 | 100 | – | – |
| 14.0 | 25–55 | 40–80 | 90–100 | – | 0–70 | 85–100 | 100 | – |
| 10.0 | 10–40 | 30–60 | 50–85 | 0–5 | 0–25 | 0–50 | 85–100 | 100 |
| 5.00 | 0–5 | 0–10 | 0–10 | – | 0–5 | 0–10 | 0–25 | 45–100 |
| 2.36 | – | – | – | – | – | – | 0–5 | 0–30 |

Table 2.14    Maximum amount to be retained on each sieve after sieving (to avoid 'blinding' of apertures)

| BS test sieve nominal aperture size (mm) | Maximum mass (kg) | | BS test sieve nominal aperture size (mm) | (µm) | Maximum mass (g) |
| | 450 mm diameter sieves | 300 mm diameter sieves | | | 200 mm diameter sieves |
|---|---|---|---|---|---|
| 50.0 | 14 | 5 | 2.36 | – | 200 |
| 37.5 | 10 | 4 | 1.18 | – | 125 |
| 20.0 | 6 | 2.5 | – | 600 | 100 |
| 14.0 | 4 | 2.0 | – | 425 | 80 |
| 10.0 | 3 | 1.5 | – | 300 | 65 |
| 5.0 | 1.5 | 0.75 | – | 150 | 50 |

Join each point by a straight line. Classify sands into zones C, M or F by reference to Table 2.12 . Note that sands may comply with more than one grading, and if they do, this should be recorded. Table 2.13 shows limits for coarse aggregates.

## Experiment 2.2    Determination of bulking of sand

*Apparatus*
A 250 ml measuring cylinder; 5 ml measuring cylinder; large beaker; wide funnel; spatula; dry sand.

*Procedure*
Weigh out 250 g of sand and transfer to the 250 ml measuring cylinder using the funnel. Tap gently to level the sand and note the volume. Transfer the sand to the beaker and add 5 ml (2 per cent) of water. Mix and again transfer to the measuring cylinder, noting the new volume after tapping. Repeat the procedure using 2 per cent increments of water until free moisture begins to separate from the sand – usually at about 20 per cent moisture content. Note that to achieve consistent results, a standardised tapping (levelling technique) is necessary.

*Results*
Measure each volume as a percentage of the original dry volume and plot graphically against moisture content. Note that even dry fine sand contains substantial quantities of air.

## Experiment 2.3   Determination of the volume of silt and clay (fines) in a sample of sand using the field-settling test

Note that the term 'fines' is used here to denote silt, clay or dust in accordance with BS 882 (1992). Note that the field-settling test is no longer a BS test.

*Apparatus*
A 250 ml measuring cylinder; common salt; sample of sand (not crushed stone sand).

*Procedure*
Prepare a 1 per cent solution of sodium chloride (common salt) in water. Two teaspoonfuls of salt to 1 litre of water will suffice.

Pour 50 ml of this solution into the 250 ml measuring cylinder. Gradually add the sand until the volume of the sand is 100 ml. Make the volume of solution up to 150 ml by adding more solution. Shake vigorously until the clayey particles are dispersed and then place on a level bench and tap gently until the surface of the sand is level. After three hours, measure the volume of silt above the sand/silt interface and express as a percentage of the volume of sand (Fig. 2.7). Refer to the text of this chapter for the acceptable limits.

## Experiment 2.4   Measurement of the moisture content of aggregates using the siphon can

*Apparatus*
Siphon can (Fig. 2.24); weighing balance of 3 kg capacity, capable of weighing to 1 g accuracy; 500 ml measuring cylinder.

*Procedure*
After ensuring that the siphon can is clean and empty, close both siphon tubes and fill until the water is above the highest point on the upper tube.

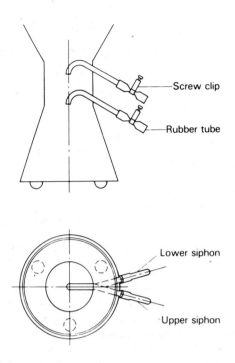

Fig. 2.24  Siphon can.

Open the upper tube and discharge the water to waste. Open the lower tube and collect the water in the measuring cylinder. The volume $V$ ml is the calibration volume for the can. Repeat and obtain an average value for $V$.

To find the moisture content of an aggregate, fill the siphon can with water and discharge using the lower siphon tube (this condition is obtained after the calibration test). Close the lower siphon tube. Weigh 2 kg of dry aggregate to 1 g accuracy and carefully add to the siphon can, stirring to remove air bubbles. Allow the stirring rod to drain and then remove it. Open the upper siphon tube and collect the water in the measuring cylinder (volume $v_b$). Open the lower siphon tube and, after draining, close both tubes. Repeat using 2 kg of the damp aggregate whose moisture content is required, obtaining the volume $v_w$ ml.

*Results*
Calculate the moisture content of the aggregate using the equations:

$$\text{Moisture content (per cent by dry mass)} = \frac{v_w - v_b}{1000 - V - v_b} \times 100$$

$$\text{Moisture content (per cent by wet mass)} = \frac{v_w - v_b}{1000 - V - v_w} \times 100$$

*Note:* These equations are based on the assumption that no water is absorbed by the aggregates; hence, that the 'dry' sample is in the saturated surface dry condition and that free water is present in the damp sample. These should apply in most cases but, where aggregates absorb water, the procedure in BS 812 should be followed.

## Experiment 2.5    Measurement of the moisture content of sand using the calcium carbide method

*Apparatus*
'Speedy' moisture meter; calcium carbide.

*Procedure*
Make sure the apparatus is clean and dry before commencing the test. Mix the main sand sample and then immediately measure out the sand sample for testing, using the scoop provided. Place this sample in the cap of the apparatus. Use the scoop to place the correct quantity of calcium carbide in the body of the instrument. Bring both parts together, keeping them horizontal to avoid mixing and tighten the cap, ensuring that the clamp screw is correctly located. Weigh the dial facing downwards, shake the instrument for three seconds and then stand, dial facing upwards. Tap to locate material in the cap and allow to stand for one minute. Repeat the shaking operation. Finally, read the instrument with the dial vertical. The reading gives the moisture content, based on wet weight directly. Empty the instrument out of doors and repeat with another sample, obtaining the average reading.

## Experiment 2.6    Determination of the slump of a concrete mix

*Apparatus*
Slump cone; hard, level, vibration-free, non-absorbent base; tamping rod; rule; trowel.

*Procedure*
Ideally the concrete should be sampled from a large batch and prepared in accordance with BS 1881, Part 101.

Alternatively, 8 litres of concrete to the specification on page 51 may be produced, assuming aggregates of this specification are available.

If neither of the above is convenient, a sample of about 8 litres may be made up using masses as follows: 3 kg cement, 5 kg sand, 8 kg coarse aggregate. The materials are mixed dry and sufficient water (about 1.5 litres) added to produce a workable concrete.

Place the clean slump cone on the base and hold firmly in position throughout the filling operation. Load concrete gently into the cone until a depth of about 100 mm is obtained. Rod 25 times with the rounded end of the tamping rod. Add another 100 mm layer and repeat, rodding through to the underlying layer. Repeat until three compacted layers are obtained and strike the top layer off level with the trowel. While still holding the cone, clean away excess concrete from around the base. Then remove the slump cone by lifting carefully in the vertical direction. Invert the cone, place next to the concrete and, using the tamping rod as a reference height, measure the slump (to the highest part of the slumped concrete) (Fig. 2.8). Record the slump in mm.

## Experiment 2.7  Determination of the workability of concrete using the compacting factor apparatus

*Apparatus*
Compacting factor apparatus (Fig. 2.9); two metal floats; vibrating table or 25 mm square tamping rod for compaction; weighing balance capable of up to 25 kg to 0.1 g accuracy; trowel; slump test tamping rod.

*Procedure*
Either obtain a sample of concrete or prepare a sample, as described in Experiment 2.6. Very wet mixes should not be used.

Weigh the collection cylinder and then place in position and cover with the two metal floats. Close both trap doors. Load the concrete gently into the upper hopper until it is well heaped. Open the upper trap door and allow the concrete to fall through into the lower hopper. If the concrete 'hangs up 'in the hopper, it may be rodded through gently with the slump test tamping rod. Remove the floats from the receiving cylinder and open the lower trap door, allowing the concrete to fill the cylinder. Cut away excess concrete by sliding the metal floats across the top of the receiving cylinder. Wipe the outside of the cylinder clean and weigh to the nearest 0.1 kg. Subtract the weight of the cylinder to obtain the weight of *partially compacted* concrete. Empty the cylinder and then refill in 50 mm layers, tamping each layer or compacting using a vibrating table to obtain *full compaction*. Weigh again and hence obtain the mass of fully compacted concrete. The compacting factor is then the ratio of the two masses:

$$\frac{\text{Mass of partially compacted concrete}}{\text{Mass of fully compacted concrete}}$$

## Experiment 2.8  Measurement of the workability of concrete using the Vebe consistometer

*Apparatus*
Vebe consistometer and vibrating table (Fig. 2.10); trowel; stopwatch.

*Procedure*
Either obtain a suitable sample of concrete or prepare a sample, as described in Experiment 2.7.

This method is most suited to dry mixes – the concrete should be barely plastic and should exhibit very little or no slump.

Fix the slump cone and outer container in position and attach the filling cone. Compact the concrete into the cone as in the slump test (Experiment 2.6). Remove the filling cone and then the slump cone and swing the Perspex disc into position, making sure the disc is free to move vertically in its guide. Rest the disc gently on the concrete. Switch on the vibrator and measure the time taken for the disc to sink, so that the concrete wets it uniformly all round. The time taken in seconds is then the consistency of the concrete in 'Vebe degrees'. It will be clear that stiffer mixes produce higher results.

## Experiment 2.9    Measurement of the compressive strength of concrete, using the cube test

It is suggested that the greatest value will be obtained from this experiment if the concrete for cube tests is produced as a series of trial mixes which will enable the effect of proportions on strength to be investigated.

*Apparatus*
Compression machine of 1000 kN or greater capacity; 100 or 150 mm cube moulds, preferably steel; vibrating table or tamping rod for compaction.

*Procedure*
First ensure that all the moulds are in good condition and that all clamps or bolts are tight. (At least five cubes per concrete sample are recommended.) Coat all internal surfaces with a *thin* film of mould oil. Mix the concrete sample thoroughly and then place in the cube moulds to a depth of approximately 50 mm. Compact this layer using a vibrating table or by hand and then add another 50 mm layer, repeating the procedure. When the moulds are full, trowel smooth and cover with an impermeable mat. Cubes must be identified, the most effective way initially being to use numbers painted on the moulds.

Leave undisturbed for 16 to 24 hours at a temperature of $20 \pm 2\,°C$ and then demould, identifying each cube with waterproof ink. Transfer to a curing tank containing clean water and cure at $20 \pm 1\,°C$ until testing.

At the required age (usually 7 or 28 days), remove the cubes from the tank and wipe away any excess water or grit. Place cubes, trowelled face sideways, centrally in the testing machine and load at the BS rate until failure. Note the maximum load and the mode of failure, which should be one of those indicated in Fig. 2.25. If there are horizontal cracks in the concrete, a tensile failure has occurred and this must be reported. Calculate the stress in $N/mm^2$ by dividing the load in N by the cross-sectional area in $mm^2$.

*Suitability of the testing machine*
Some machines in use have been found to be unsatisfactory for cube testing. Machines should comply with BS 1881, Part 115 (1986), and periodic

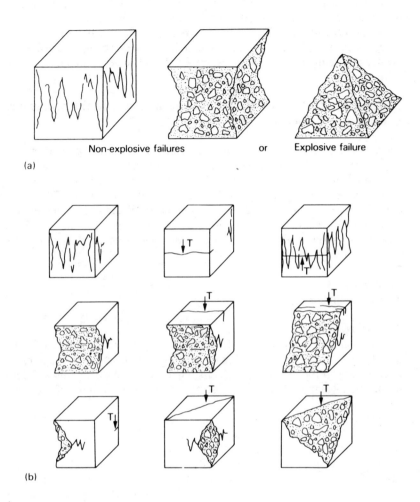

Non-explosive failures    or    Explosive failure

(a)

(b)

Fig. 2.25   Modes of failure of concrete cubes: (a) normal, (b) abnormal – tensile cracks marked T are indicative of incorrect testing or faulty machine.

checks are necessary that the upper ball seating works properly and that bearing surfaces are not excessively worn. The ball seating forms part of a sphere with its centre at the face of the upper platen. It is designed to rotate during initial loading to align accurately with the upper face of each concrete cube. Thereafter it should lock by friction in order to avoid premature crushing of any weak portion of the concrete cube (for example, the trowelled face). Where failures of the form of Fig. 2.25(b) occur they may indicate problems with the ball seating. A good way of checking the machine generally is to use the 'reference test service' operated by a number of organisations with suitable machines. Pairs of 'identical' cubes are made and tested on each machine to reveal shortcomings in the machine in question; if the machine is in good condition test results should be very similar.

## Experiment 2.10   The effect of mix proportions on the properties of fresh and hardened concrete

*Note:* A comprehensive investigation into the effect of mix proportions on the properties of concrete will require the production of a considerable number of mixes. Where time permits, a series of nine mixes might be made, each mix being designed by the DoE method of mix design. Table 2.15 indicates nine such mixes that might be produced. Where time is limited, the study might be confined to three mixes, for example Nos 1, 5 and 9. (These three mixes should help students to overcome the common fallacy that dry mixes are always stronger.)

*Apparatus*
As for Experiments 2.6 and 2.9

*Procedure*
Select a number of mixes from Table 2.15 and produce batch quantities by use of the form of Fig. 2.11, using the same size, type and size or aggregate in each case. (Produce a target mean strength on the basis that no previous results are available.) For the purpose of the slump test, 8 litres of each mix will be required.

Measure the free-moisture contents of the aggregates and for each mix correct the batch quantities, as described on page 33 .

For each concrete mix proceed as follows:

Mix the materials dry and then add a quantity of water slightly less than the required batch quantity. Try to estimate the likely slump and then continue to add water until what is judged to be the correct slump is obtained. Then carry out the slump test. Note that workability tests should, if

Table 2.15   Suggested mix numbers for trial mixes

| 28 day characteristic strength ($N/mm^2$) | Slump (mm) | | |
|---|---|---|---|
| | 0–30 | 30–60 | 60–180 |
| 15 | 1 | 2 | 3 |
| 25 | 4 | 5 | 6 |
| 40 | 7 | 8 | 9 |

possible, be carried out at a standard time (usually 6 min) after adding the mixing water. If the workability is outside the acceptable range, another mix should, strictly speaking, be produced.

The cohesiveness of the concrete should also be measured – produce a small heap of the concrete and trowel to a thickness of about 50 mm. A mix of normal cohesion should trowel to a smooth finish fairly easily. Mixes of high or low cohesion will be very easy or very difficult, respectively, to trowel to a smooth finish.

Make up at least five cubes from the batch, curing and testing as described in Experiment 2.9. All results should be recorded and compared with the original specification. The variability within each set of five cubes should be measured by calculation of the standard deviation and compared with values indicated on page 55. (Note that five results are not really sufficient for a full analysis and that laboratory-produced concrete should have a relatively low standard deviation.) The cost of the concrete varies directly with the cement content; note the mix that is most expensive on this basis.

## Questions

1. List the four chief components in Portland cement. State which of these:

    (a) gives off most heat during hydration
    (b) is present in the largest quantity of OPC
    (c) is vulnerable to sulphates
    (d) is responsible for the grey colour of OPC

2. State what is meant by a *hydraulic* cement.
3. Indicate two ways in which the manufacturer of a cement could increase its rate of hardening using given raw materials. Give typical applications of rapid-hardening cements.
4. State the cause of sulphate attack in concrete. Indicate two ways in which the likelihood of this attack can be reduced.
5. Name the two principal compounds produced on hydration of ordinary Portland cement. Give one characteristic property of reinforced concrete which depends on each.
6. Explain the purpose of the following tests for cement:

    (a) setting time
    (b) soundness
    (c) mortar cube strength

7. Explain the factors which affect the choice of the maximum size of aggregate for concrete and indicate what maximum size of aggregate would be suitable for the following purposes:

(a) a normally reinforced concrete column of section 600 mm × 400 mm
(b) a 40 mm floor screed
(c) a concrete road base of thickness 200 mm

8. A sieve analysis of 250 g of sand gives the following results:

| Sieve size | Mass retained (g) |
| --- | --- |
| 10 mm | 0 |
| 5 | 5 |
| 2.36 | 31 |
| 1.18 | 38 |
| 600 µm | 38 |
| 300 | 79 |
| 150 | 51 |
| Passing 150 µm | 8 |

Classify the aggregate by means of Table 2.12.

9. Explain what is meant by 'bulking' of aggregates. Indicate why the extent of bulking of sand depends on moisture content.

10. A sample of damp aggregate weighing 2.35 kg is dried by hair dryer until it just reaches the free-running (saturated surface dry ) condition. It is then found to weigh 2.24 kg. After drying in the oven at 110 °C to constant mass, it is found to weigh 2.15 kg. Calculate:

(a) the free-water content
(b) the total water content

based on dry mass.

11. Compare the following tests for the workability of concrete:

(a) slump test
(b) compacting factor test
(c) Vebe test

A sample of concrete was divided into three and results for the slump, compacting factor test and Vebe tests obtained. They were as follows:

slump, 5 mm; compacting factor, 0.93; Vebe time, 18 s.

Assuming that only one result was in error, which result is incorrect? (Refer to Table 2.6.)

12. Explain what is meant by the term 'cohesion' in relation to fresh concrete. Indicate the problems that might arise in concretes of

(a) very low cohesion
(b) very high cohesion

13. Use Table 2.8 and Fig. 2.13 to determine the water/cement ratios required in mixes to the following specifications:

| Cement/ aggregate | Target mean strength at stated age (N/mm$^2$) | | |
|---|---|---|---|
| | 3 days | 7 days | 28 days |
| 1  OPC/uncrushed | (a)  25 | (b)  22 | (c)  40 |
| 2  OPC/crushed | (a)  20 | (b)  40 | (c)  60 |
| 3  RHPC/crushed | (a)  30 | (b)  35 | (c)  60 |

**14.** Use the DoE method of mix design to produce batch quantities for 50 litres of concrete to the following specification:

| | |
|---|---|
| Cement | Ordinary Portland |
| Aggregates | Sand – uncrushed, relative density 2.6; 59 per cent passes the 600 µm sieve |
| | Coarse – 20 mm, uncrushed, relative density 2.6. |
| Compressive strength | 35 N/mm$^2$ at 28 days (5 per cent defectives), no previous results available |
| Workability | 30–60 mm slump |

**15.** Use the DoE method of mix design to produce batch quantities for 10 litres of concrete to the following specification:

| | |
|---|---|
| Cement | Rapid-hardening Portland |
| Aggregates | Sand – uncrushed, relative density 2.7; 75 per cent passes the 600 µm sieve |
| | Coarse – uncrushed, relative density 2.7 maximum size 40 mm |
| Minimum cement content | 350 kg/m$^3$ concrete |
| Compressive strength | 35 N/mm$^2$ at 28 days (5 per cent defectives allowed). Standard deviation 5 N/mm$^2$ |
| Workability | 10–30 mm slump. |

**16.** Concrete is required to be produced with a characteristic strength of 25 N/mm$^2$ with 2 per cent failures allowed. The first 20 cube results are as follows:

| | | | | |
|---|---|---|---|---|
| 24.2 | 23.9 | 28.1 | 20.2 | 29.4 |
| 32.5 | 19.8 | 24.2 | 26.5 | 23.2 |
| 30.5 | 27.5 | 24.1 | 26.5 | 29.1 |
| 20.1 | 24.5 | 25.1 | 22.9 | 21.2 |

Calculate the standard deviation using the formula given on page 54 and hence indicate whether the required characteristic strength is being reached.

**17.** Concrete for a particular contract is to be produced with a target mean strength of 33 N/mm$^2$ and the assumed standard deviation is 4.5 N/mm$^2$. The first 50 results are as follows (in N/mm$^2$):

| 26.2 | 28.0 | 34.8 | 39.0 | 40.0 | 28.2 | 30.0 | 38.6 | 37.0 | 33.0 |
|------|------|------|------|------|------|------|------|------|------|
| 34.6 | 33.6 | 35.0 | 29.6 | 28.8 | 30.0 | 26.0 | 24.4 | 29.2 | 36.0 |
| 40.0 | 30.2 | 33.4 | 36.8 | 34.9 | 26.0 | 26.8 | 30.6 | 37.6 | 31.6 |
| 30.2 | 28.4 | 37.2 | 35.6 | 30.6 | 40.6 | 37.2 | 30.4 | 33.4 | 34.8 |
| 38.6 | 33.8 | 33.6 | 28.8 | 32.8 | 34.8 | 39.0 | 28.4 | 34.6 | 32.8 |

Draw a control chart of the type shown in Fig. 2.18, including the target mean strength and upper and lower control lines for 5 per cent and 10 per cent of *individual* results. Hence suggest whether the characteristic strength requirement of 25 N/mm$^2$ with 5 per cent permissible failures is being satisfied.

18. Concrete is required to have a characteristic strength of 30 N/mm$^2$ at 28 days with 5 per cent permissible failures. Calculate the water/cement ratio which will be needed if OPC and uncrushed aggregates are used. Assume a standard deviation of 5.5 N/mm$^2$.

19. Indicate three alternative reasons for using:

    (a) plasticisers
    (b) retarders

    in concrete.

## References

*Design of Normal Concrete Mixes*, Department of the Environment, HMSO, 1988.

## British Standards

BS 12: 1991, *Portland cements.*
BS 146: 1973, *Portland blast furnace cement.*
BS 812, *Testing aggregates.*
    Part 101: 1984, *Guide to sampling and testing aggregates.*
    Part 102: 1989, *Methods for sampling aggregates.*
    Part 102: 1989, *Methods for sampling.*
    Part 103: *Method for determination of particle size distribution.* Section 103.1: 1985, *Sieve tests.* Section 103.2: 1989, *Sedimentation test.*
    Part 109: *Methods of determination of moisture content.*
BS 882: 1992, *Specification for aggregate from natural sources for concrete.*
BS 1370: 1979, *Low-heat Portland cement.*
BS 1881, *Testing concrete.*
    Part 101: 1983, *Methods of sampling fresh concrete on site.*
    Part 102: 1983, *Method for determination of slump.*
    Part 103: 1983, *Method of determination of compacting factor.*

Part 104: 1983, *Method of determination of Vebe time.*
Part 108: 1983, *Method of making test cubes from fresh concrete.*
Part 111: 1983, *Method of curing of test specimens (20 °C method).*
Part 115: 1986, *Specification for compression test machines for concrete.*
Part 116: 1983, *Method of determination of compressive strength of concrete cubes.*
BS 4027: 1991, *Specification for sulphate-resisting Portland cement.*
BS 5075, Part 1: 1982, *Accelerating admixtures, retarding admixtures and water-reducing admixtures.*
BS 5075, Part 2: 1982, *Air-entraining admixtures.*
BS 6588: 1991, *Specification for Portland pulverised-fuel ash cements.*
BS 6610: 1991, *Specification for pozzolanic pulverised-fuel ash cements.*
BS 8110, *Structural use of concrete*
Part 1: 1985, *Code of Practice for design and construction.*

# Chapter 3

# Bricks, building blocks and mortars

## Bricks

Bricks may broadly be described as building units which are easily handled with one hand. By far the most widely used size at present is the single standard metric brick of actual size 215 × 102.5 × 65 mm. The relationship between these dimensions is illustrated in Fig. 3.1. It gives great versatility to the use of standard metric bricks.

For the past 100 years, clay bricks have dominated the UK market as building units but more recently other brick types and concrete blocks have provided strong competition, especially since thermal insulation regulations were tightened and lightweight concrete blocks became available. In terms of availability, clay bricks continue to be the most important building unit, combining excellent durability (when they are selected and used correctly) with, in the case of facing bricks, lasting aesthetic properties.

## Clay bricks

These are made by pressing a prepared clay sample into a mould, extracting the formed unit immediately and then heating it in order to sinter (partially vitrify) the clay.

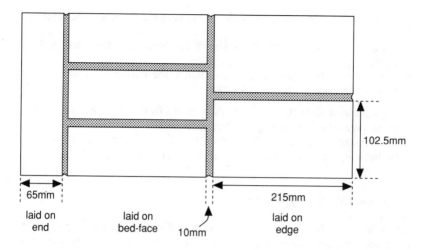

Fig. 3.1   Relationships between the dimensions of a standard metric brick.

Many different types of brick may be produced, depending on the type of clay used, the moulding process and the firing process.

There are three basic subdivisions of type.

1.  *Common bricks.* These are ordinary bricks which are not designed to provide good finished appearance or high strength. They are therefore in general the cheapest bricks available.
2.  *Facing bricks.* These are designed to give attractive appearance, hence they are free from imperfections such as cracks.
3.  *Engineering bricks.* These are designed primarily for strength and durability. They are usually of high density and well fired.

Bricks may also be classified in other ways, for example by their frost resistance or their soluble salt content.

Many clay bricks are designated by their place of manufacture, colour or surface texture. Examples:

*   *Fletton* – a common brick manufactured from Oxford clay, originally in Fletton, near Peterborough.
*   *Staffordshire blue* – an engineering quality brick produced from clay which results in a characteristic blue colour.
*   *Dorking stock* – the term 'stock', originally a piece of wood used in the moulding process, now denotes a brick characteristic of a certain region.

## Indentations and perforations in bricks

Indentations (frogs) and perforations (cylindrical holes passing through the thickness of the brick) may be provided for one or more of the following reasons:

- They assist in forming a strong bond between the brick and the remainder of the structure.
- They reduce the effective thickness of the brick and hence reduce firing time.
- They reduce the material cost and hence the overall cost of the brick without serious *in-situ* strength loss.

For greatest strength, bricks with a single frog should be laid frog-up, since this ensures that the frog is filled with mortar.

Perforated bricks generally contain less than 25 per cent of voids and these behave, in a structural sense, as if they were solid .

## Manufacture of clay bricks

There are four basic stages in brick manufacture, though many of the operations are interdependent – a particular brick will follow through these stages in a way designed specifically to suit the raw material used and the final product.

### Clay preparation

After digging out, the clay is prepared by crushing and/or grinding and mixing until it is of a uniform consistency. Water may be added to increase plasticity (a process known as 'tempering') and in some cases chemicals may be added for specific purposes – for example, barium carbonate which reacts with soluble salts producing an insoluble product.

### Moulding

The moulding technique is designed to suit the moisture content of the clay, the following methods being described in order of increasing moisture content.

*Semi-dry process.* This process, which is used for the manufacture of Fletton bricks, utilises a moisture content in the region of 10 per cent. The ground and screened material has a granular consistency which is still evident in fractured surfaces of the fired brick. The material is pressed into the mould in up to four stages. The faces of the brick may, after pressing, be textured or sandfaced.

*Stiff plastic process.* This utilises clays which are tempered to a moisture content of about 15 per cent. A stiff plastic consistency is obtained, the clay being extruded and then compacted into a mould under high pressure. Many Engineering bricks are made in this way, the clay for these containing a relatively large quantity of iron oxide which helps promote fusion during firing.

*The wire-cut process.* The clay is tempered to about 20 per cent moisture content and must be processed to form a homogeneous material. This is extruded to a size which allows for drying and firing shrinkage and units are cut to the correct thickness by tensioned wires. Perforated bricks are made in this way, the perforations being formed during the extrusion process. Wire-cut bricks are easily recognised by the perforations or 'drag marks' caused by the dragging of small clay particles under the wire.

*Soft mud process.* As the name suggests, this process utilises a clay, normally from shallow surface deposits, in a very soft condition, the moisture content being as high as 30 per cent. Breeze or town ash may be added to provide combustible material to assist firing or improve appearance. The clay is pressed into moulds which are sanded to prevent sticking. The green bricks are very soft and must be handled carefully prior to drying.

Hand-made bricks are produced by a similar process, except that 'clots' of clay are thrown by hand into sanded moulds. This produces a characteristic surface texture (Fig. 3.2), which has great aesthetic appeal and is the reason for the continued production, on a limited scale, of hand-made bricks.

### Drying

This must be carried out prior to firing when bricks are made from clay of relatively high moisture content. Drying enables such bricks to be stacked higher in the kiln without lower bricks becoming distorted by the weight of the bricks above them. Drying also enables the firing temperature to be increased more rapidly without such problems as bloating, which may result when gases or vapour are trapped within the brick. Drying is carried out in chambers, the temperature being increased and the relative humidity progressively decreased as bricks lose moisture. The process normally takes a number of days, higher moisture-content bricks requiring a greater time. Wire-cut bricks and those produced by the soft mud process must be dried prior to firing.

### Firing

The object of firing is to cause localised melting (sintering) of the clay, which increases strength and decreases the soluble salt content without loss of shape of the clay unit. The main constituents of the clay – silica and alumina – do not melt, since their melting points are very high; they are

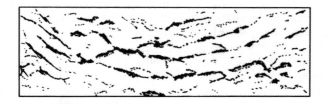

Fig. 3.2   Characteristic surface texture of a hand-made clay brick. Each brick is slightly different.

merely fused together by the lower melting point minerals such as metallic oxides and lime. The main stages of firing are:

| | |
|---|---|
| 100 °C | evaporation of free water |
| 400 °C | burning of carbonaceous matter |
| 900–1000 °C | sintering of clay |

The latter stages of firing may be assisted by fuels which are either present naturally in the clay or added during processing – one of the factors contributing to the low cost of Fletton bricks is the presence of carbonaceous material, which is largely responsible for the heat supplied in the later stages of firing.

The control of the rate of increase of temperature and the maximum temperature is most important in order to produce bricks of satisfactory strength and quality; in particular, too rapid firing will cause bloating and overburning of external layers, while too low a temperature seriously impairs strength and durability. Stronger bricks, such as Engineering bricks, are normally fired at higher temperatures.

The following are the main processes:

*Clamps.* Bricks are stacked in large special formations on a layer of breeze, though the bricks also contain some fuel. The breeze base is ignited and the fire spreads slowly through the stack, which contracts as the bricks shrink on firing. The process may take up to one month to complete and the fired product is very variable, many underburnt and overburnt bricks being obtained. After firing, the bricks are sorted and marketed for various applications. Well-fired bricks are extremely attractive with local colour variations caused by temperature differences and points where fuel ignition occurred. The use of clamps has now decreased, owing to the difficulty in controlling these kilns and the high wastage involved.

*Continuous kilns.* These are based on the Hoffman kiln and comprise a closed circuit of about 14 chambers arranged in two parallel rows with curved ends (Fig. 3.3). Divisions between the chambers are made from strong paper sealed with clay and, by means of flues, the fire is directed to each chamber in turn. Drying is carried out prior to the main firing process and is achieved by warm air obtained from fired bricks during cooling. The kilns are described as continuous, since the fire is not extinguished – it is simply diverted from one kiln to the next, the cycle taking about one week. Coal was traditionally used but firing now may be by oil or gas. These kilns are very widely used for brick production.

*Tunnel kilns.* These are the most recently introduced kilns and they can reduce firing time to little over one day. Units are specially stacked onto large trolleys incorporating a heat-resistant loading platform. The trolleys are then pushed end-to-end into a straight tunnel with a waist that fits the loading platform closely. The bricks pass successively through drying, firing and cooling zones, firing normally being by oil or gas. The process provides a

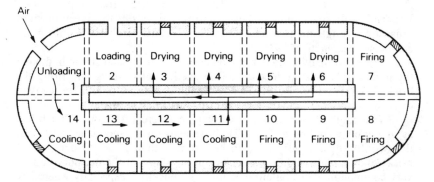

Fig. 3.3 Continuous kiln showing sequence of zones and movement of air for drying purposes.

high degree of control over temperature, so that the process is suited to the production of high-strength, dimensionally accurate bricks. Perforated bricks are often fired in this way.

## Properties of clay bricks and their measurement

### *Strength*

Since bricks are invariably used in compression, the standard method of test for strength involves crushing the bricks, the direction of loading being the same as that which is to be applied in practice – normally perpendicular to their largest face. Attention is drawn to the following aspects of the BS test (see Experiment 3.1).

- The treatment of frogs or perforations should be designed to simulate their behaviour in practice – hence the perforations in perforated bricks should not be filled prior to testing and frogs should only be filled if the bricks are to be used frog-up. When frogs are filled, the strength of the mortar used should be in the range 28 to $42\,N/mm^2$ at the time of testing.
- Strength varies with moisture content and, since the saturated state is the easiest state to reproduce, bricks should be saturated for 24 hours (or by vacuum or boiling) before testing.
- In order to obtain representative results, 10 bricks should be tested, these being carefully sampled either during unloading or from a stack.
- Bricks are tested between 4mm thick plywood sheets which reduce stress concentrations resulting from irregularities in the brick surfaces.
- The strength of the brick is:

$$\frac{\text{Maximum load}}{\text{Area of smaller bed face}}$$

(the two bed faces may not always be identical in size).

Bricks may be designated as class 1, 2, 3, 4, 5, 7, 10 or 15, according to their average compressive strength. These classes derive from formerly used Imperial units, the numbers being multiplied by 6.9 to give minimum crushing strength in $N/mm^2$. Hence, a class 10 brick has a minimum compressive strength of $69\,N/mm^2$.

There is normally good correlation between density and strength, though the precise relationship depends on the method of forming and firing the brick. Underfired bricks would have much reduced strength for a given density, as would bricks which are damaged by firing too rapidly.

The strength of *brickwork* is usually quite different from that of the brick, as measured according to BS 3921. It depends on the mortar used (though this would not normally need to be stronger than 10 to 20 per cent of brickwork strength) and, in particular, on the shape of the masonry unit. A very common mode of failure in walls is by buckling, especially when they are tall and slender with little lateral restraint.

## Water absorption

Water absorption may be an important property of clay bricks, since bricks that have very low absorption are invariably of high durability (though the converse is not always true).

In the BS water absorption test (Experiment 3.2), oven-dried bricks are either boiled for 5 h or subjected to a vacuum test in order to ensure maximum possible penetration of water (even in these tests there will normally be some voids which remain unfilled). An alternative method for control purposes only is to soak the brick in water for 24 h, though this gives a lower result than the first two tests.

The water absorption is:

$$\frac{\text{Mass of water absorbed}}{\text{Mass of oven-dried brick}}$$

## Efflorescence

This is the name given to the build-up of white surface deposits on drying. It results from dissolved salts in the brick and quite commonly spoils the appearance of new brickwork.

In the test for efflorescence, bricks are saturated with distilled water in order to dissolve any salts present and then allowed to dry such that salts are carried to one exposed face (Experiment 3.3). The apparatus is shown in Fig. 3.4.

The efflorescence is assessed as follows:

Nil          No perceptible deposit of efflorescence

Slight       Not more than 10 per cent of area covered with thin deposit of salts but unaccompanied by powdering or flaking of the surface.

Moderate     A heavier deposit than slight covering up to 50 per cent of the area of the face but unaccompanied by powdering or flaking of the surface

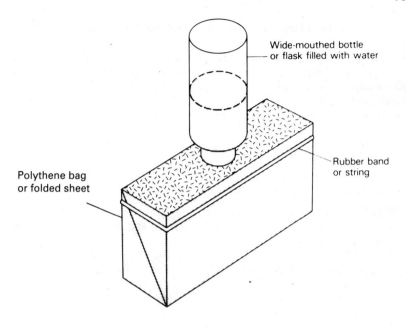

Wide-mouthed bottle
or flask filled with water

Rubber band
or string

Polythene bag
or folded sheet

Fig. 3.4   Apparatus for efflorescence test.

Heavy        A deposit of salts covering more than 50 per cent of the area of
             the face and/or powdering or flaking of the surface

When heavy efflorescence is obtained as defined above, the consignment is
considered not to comply with the standard.

*Soluble salt content*
Salts can be harmful in causing efflorescence but also in leading to problems
such as sulphate attack in the mortar. Bricks are classified as of low (L)
soluble salt content if the percentages of the following ions do not exceed the
levels stated:

|  |  |
|---|---|
| Sulphate | 0.5 per cent |
| Calcium | 0.3 per cent |
| Magnesium | 0.03 per cent |
| Potassium | 0.03 per cent |
| Sodium | 0.03 per cent |

The figure for the calcium ion is larger than for the other metallic ions
because it has lower solubility. Bricks having higher values would be
classified as 'N', though levels are still effectively limited by the
efflorescence requirement.

## British Standard requirements for Engineering bricks

Engineering bricks have specific requirements relating to absorption and strength in addition to those such as dimensional tolerances and efflorescence requirements which apply to other types. Table 3.1 indicates these requirements for Engineering class A and class B bricks. The water absorption values are based on the 5 h boiling test.

### Frost resistance

Frost resistance of clay bricks is designated as follows:

- *Frost resistant* (F). Bricks durable in all building situations including those where they are in a saturated condition and subjected to repeated freezing and thawing.
- *Moderately frost resistant* (M). Bricks durable except when in a saturated condition and subjected to repeated freezing and thawing.
- *Not frost resistant* (O). Bricks liable to be damaged by freezing and thawing if not protected as recommended in BS 5628, Part 3, during construction and afterwards, for example, by an impermeable cladding. Such units may be suitable for internal use.

### Durability of clay brickwork

The durability of clay brickwork is much more likely to be a problem than its strength, since in most situations clay bricks are very much stronger than is required structurally.

Virtually all durability problems are associated with moisture penetration and it is therefore of paramount importance that bricks be suited to the degree of exposure likely to be found. In any one structure the most exposed position should be identified and, assuming that the same brick type is to be used throughout, the brick selected should be adequate for this degree of exposure.

The causes and mechanism of the three main modes of deterioration are now considered.

Table 3.1    Requirements for Engineering bricks

| Class | Compressive strength (N/mm²) | Water absorption (%) |
|---|---|---|
| Engineering A | 70 | 4.5 |
| Engineering B | 50 | 7.0 |

*Frost damage* (see Experiment 3.4)

This is associated with the crystallisation (freezing of water) below the surface of the bricks. Hence, for this to occur the bricks must be fairly porous in order to admit water. There is strong evidence to suggest that water contained in bricks moves into cracks or large voids or towards the surface as the temperature is lowered, damage occurring as ice crystals grow in any enclosed positions. Figure 3.5 shows ice crystals being extruded from pores and cracks in the brickwork. Damage is progressive: a few freezing cycles rarely cause visible damage and it will normally take several years for frost damage to become apparent. The susceptiblity of bricks to frost damage clearly increases as their water content rises, though some highly porous bricks are found to be frost-resistant even when saturated. Investigations have shown that it is the size distribution of pores in a brick rather than total porosity which determines frost resistance; bricks containing coarse pore structure are generally more frost-resistant than those with a fine pore structure. Figure 3.6 shows fractured surfaces of a Fletton (type 'O') brick and an Otterham (type 'F') brick magnified about 300 times using a scanning electron microscope. Although the two brick types have similar absorption values of about 20 per cent, the degree of vitrification is seen to be larger in the case of the Otterham brick, resulting in fewer fine spaces within the brick

Fig. 3.5   Ice extrusions from pores/cracks in brickwork, subject to freezing when saturated.

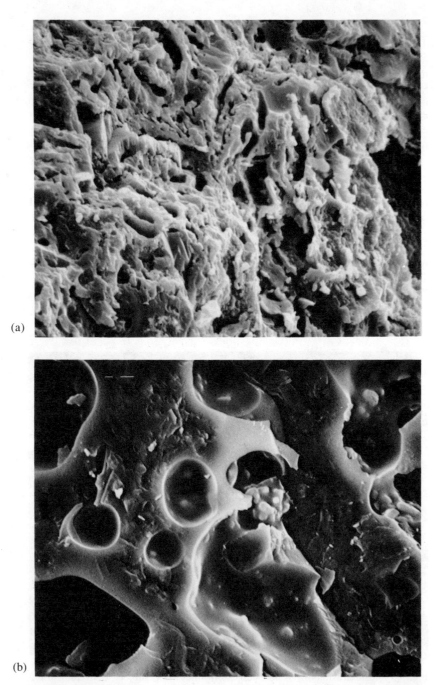

(a)

(b)

Fig. 3.6 Scanning electron microscope photographs of fracture surfaces in (a) Fletton brick (type 'O') and (b) Otterham brick (type 'F').

and therefore a more frost-resistant product. However, since pore size distribution is difficult to measure, frost resistance is best assessed from long-term experience. Perhaps surprisingly, some stock bricks having quite low strength and high water absorption are found to be frost-resistant, presumably because they contain a coarse pore structure.

As a general rule, the resistance of brickwork to frost will be greatest when all possible means are taken to prevent moisture penetration. In the case of types 'M' and 'O' bricks, the highest risk may be present during construction; it is important to keep bricks dry and to avoid high moisture contents in partially complete structures which do not have the protection of a roof, especially during cold winter spells.

*Crystallisation damage* (see Experiment 3.5)
This is associated with the crystallisation of salts beneath the brick surface. The extent of damage is directly related to the soluble salt content of bricks, hence the best way of avoiding damage is to use type 'L' bricks. Salts from other sources may also contaminate brickwork. Such sources include poor-quality sand in the mortar, contaminated ground water and wind-borne spray from the sea. Nitrates or chlorides are more common than sulphates, though they all have largely the same effect. Some salts may even be released from the cement, though the effect of these should be short-lived.

One problem of salt crystallisation is that it tends to occur locally at boundaries between damp areas of brickwork, such as parapet walls and adjacent drier areas. Salts in solution are drawn to these drier areas and deposits therefore build up, causing local damage. Many salts are hygroscopic (i.e. they absorb water), and may tend to perpetuate dampness where they occur. Hence, in order to avoid these problems:

- Use well-fired bricks or, in situations where dampness is likely, type 'L' bricks.
- Use clean materials and avoid contact between brickwork and ground water or soil.
- Design and detail brickwork in such a way as to minimise moisture penetration.

*Sulphate attack*
This occurs in cement mortars, especially those in which the cement contains appreciable quantities of tricalcium aluminate, though the sulphates responsible often originate in the bricks. Sulphate attack will only occur in damp situations and it results in expansion and eventual disruption of the brickwork. Figure 3.7 shows disruption of the mortar joints in a clay brick retaining wall. Avoidance of sulphate attack is achieved by the same precautions as given under the section on crystallisation damage, though the use of sulphate-resisting cement in the mortar will reduce the severity of attack where these precautions cannot be met fully, such as in earth-retaining walls.

Fig. 3.7 Sulphate attack in clay brickwork retaining wall. The sulphates originated both in the bricks and in the ground (the wall has no dpc).

## Calcium silicate (sand/lime) bricks (BS 187)

These bricks are made by blending together finely ground sand or flint and lime in the approximate ratio 10:1. The semi-dry mixture is compacted into moulds and then autoclaved, using high-pressure steam for several hours. A surface reaction occurs between the sand and lime, producing calcium silicate hydrates which 'glue' the sand particles into a solid mass. The main properties of calcium silicate bricks are:

- A high degree of regularity, with a choice of surface texture ranging from smooth to 'rustic'.
- A wide range of colours, produced using pigments.
- Very low soluble salt content, so that efflorescence is not a problem.
- Fairly high moisture movement.
- Compressive strength in the range 7–50 N/mm$^2$ (strength classes are employed in BS 187).
- Good overall durability in clean atmospheres, though they may deteriorate slowly in polluted sulphur-containing atmospheres.

Selection should, as with clay bricks, be according to purpose but, unlike clay bricks, strength is the best guide to durability. Uses are as follows:

| | |
|---|---|
| Class 2 | Protected external applications |
| Class 3 | Free-standing external, or below dpc |
| Class 4 or greater | Copings and retaining walls |

BS 187 requires that calcium silicate bricks have a *shrinkage* not exceeding $400 \times 10^{-6}$; this is of the same order or magnitude as the initial *expansion* of clay bricks. On account of their increased moisture movement, joints in calcium silicate brickwork are recommended every 7 m, approximately, and they should not be laid wet. Since movement characteristics are different, they should not be bonded directly to clay brickwork. The use of calcium silicate bricks has increased considerably in recent years, due at least in part to the fact that they are now available in a variety of colours, such as pastel shades of red and blue, as well as the original off-white colour.

## Concrete bricks (BS 6073)

These are of relatively recent introduction, comprising well-compacted, low-workability concrete mixes of appropriate aggregate size, leading to products of high strength and durability. Colours and textures can be obtained which can give a final appearance very similar to that of clay bricks. They are free from efflorescence, tending to shrink slightly in dry situations. Owing to different movement characteristics, they should not be bonded to other brick types. With a limited number of production plants currently operating, transport costs may be high except in the regions where they are produced. There may also be problems of obtaining a good colour match if, at a later stage, extension or modification of buildings on which such bricks were used is undertaken.

## Building blocks

These may be defined as building units having at least one dimension in excess of the maximum specified for bricks (see 'Bricks'). They are designed to be bedded in mortar to produce walling or to be used as filler blocks in reinforced concrete floors. Blockwork has developed into a major form of construction for the following reasons:

1. Rates of production during construction are substantially greater with blocks than with brick-size units – for example, a 100 mm thick block of size $440 \times 215$ mm is equivalent to approximately six standard bricks.

2. A great variety of sizes and types is available to suit purposes ranging from structural use to lightweight partitions.
3. Modern factory production methods ensure consistent and reliable performance.
4. High-quality surface finishes are obtainable, obviating the need for rendering or plaster coatings in many cases.

## Clay building blocks

Clay blocks are generally hollow units (formed by extrusion), since the voids result in:

- reduced firing times
- reduced weight
- increased handleability
- increased thermal insulation

Figure 3.8 shows various types of block available. Since blocks are made from well-sintered clay, the material is dense, brittle and difficult to cut. Some varieties, however, contain slots which assist in cutting.

Clay wall and partition blocks are keyed for rendering or plastering, as they are not intended to be used as finished facing materials.

## Concrete blocks

Concrete blocks are covered by BS 6073, which also covers concrete bricks. They may be solid, hollow, or cellular.

- *Solid blocks* – largely voidless but may have grooves or holes to reduce weight or facilitate handling. These must not exceed 25 per cent of the gross volume of the block.
- *Hollow blocks* – these voids passing right through. They can be 'shell-bedded' – that is, the mortar is laid in two strips adjacent to each face, so that there is no continuous capillary path for moisture through the bed. Loadbearing capacity is, of course, reduced when blocks are laid in this way. The strength of hollow blockwork can be increased by filling the cavities with concrete, especially if reinforcement is included. Sound insulation is also improved in this way.
- *Cellular blocks* – a special type of hollow block in which the cavities are closed at one end. The solid edge would normally be laid upwards and, in the case of thin blocks, this makes it easier to produce an effective bed joint.

A great variety of aggregate types is used in the production of concrete blocks – these include natural aggregates, air-cooled blast-furnace slag, furnace clinker, foamed, expanded or granulated blast-furnace slag, bottom ash from boilers and milled softwood chips. The density and strength of the resultant product vary accordingly and they are also influenced by the manufacturing technique.

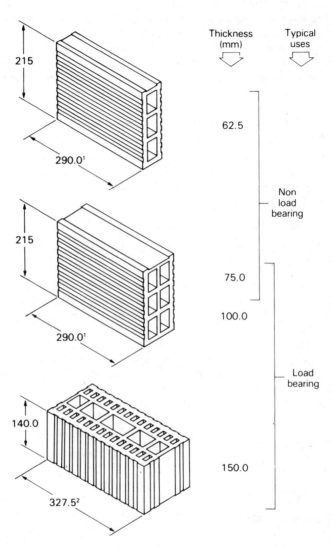

Thickness (mm)

Typical uses

215

290.0¹

62.5

Non load bearing

215

290.0¹

75.0

100.0

Load bearing

140.0

327.5²

150.0

¹ Special bonding lengths are available.
² These blocks can be easily cut to length.

Fig. 3.8   Clay building blocks. (From Mitchell's Building Series: Materials, Longman.)

Denser types of block can be used for loadbearing purposes and decorative or textured facings may be applied.

Lightweight types are now in particular demand, due to the need for high thermal insulation walling. Many types include pulverised fuel ash and may be aerated and/or autoclaved. Higher insulation varieties of autoclaved aerated blocks are easily cut with hand saws and can be used without cavities to produce walling to current thermal insulation requirements. (The waterproofing function would be achieved by application of a suitable external rendering.)

BS 6073 does not stipulate strength classes but gives minimum strengths for all block types. These are:

- Thickness not less than 75 mm: Average strength of 10 blocks not less than 2.8 N/mm$^2$. No individual block less than 80 per cent of this value.
- Thickness less than 75 mm: Average transverse strength of 5 blocks not less than 0.65 N/mm$^2$.

BS 6073 also stipulates drying shrinkage, which should be:

- not more than 0.06 per cent (except autoclaved aerated blocks)
- not more than 0.09 per cent (autoclaved aerated blocks)

It is important, on account of their high shrinkage, that autoclaved aerated blocks are not saturated prior to laying, otherwise shrinkage cracking may result. Handleability is also much impaired when wet due to their greatly increased weight in this condition.

No guidance regarding durability of blocks is given in BS 6073. The principles given in Chapter 2 on concrete apply, though it may be added that some open-textured or autoclaved blocks are often more resistant to frost than their strength would suggest. This may be due to the type of void present and the fact that saturation is rare. Nevertheless, where severe exposure or pollution is likely, blocks of average compressive strength not less than 7 N/mm$^2$ should be specified. All types can, of course, be protected by rendering.

## Masonry mortars

Mortars may be defined as mixtures of sand, a binder such as lime or cement, and water. Their prime function in masonry is to take up tolerances between building units such as bricks or blocks, though they also have to satisfy some or all of the following requirements.

(a) They should impart sufficient strength to the complete unit.
(b) They should permit movement (unless this is negligible or joints are provided). When movement occurs within a well-constructed masonry

unit, it should take place in the form of microcracks within the mortar rather than cracking of the bricks or blocks.

(c) They should be durable, that is, resist frost or other forms of environmental attack.

(d) They should resist penetration of water through the unit.

(e) They should contribute to the aesthetic appearance of the wall.

To permit effective use, mortars should be workable, yet cohesive in the fresh state; bricks or blocks should need only minimal effort to bed them in the correct position.

The functions (a), (c) and (d) above are best provided by a strong mortar, though the wall as a whole will be in the region of five times stronger than the mortar on account of the very small mortar thicknesses used – typically 10 mm. Function (b) is best satisfied by weak mortars which crack readily if movement occurs. This leads to the working rule:

*'the mortar must not be stronger than the units it is bonding'*

Mortar mixes are classified in BS 5628, Part 3 (Table 3.2). For high-strength units such as engineering bricks, type (i) can be used, the cement content being almost sufficient to produce a cohesive mortar without additional plasticisers. These mortars will give great strength and durability and can be used in situations of high exposure and high stress, though separate movement provision may be required. Some 'softer' sands may not give adequate durability with type (i) mortars, since they are too fine and lead to high water requirements. Coarser sands of BS 1200 or finer concreting sands of BS 882 should be suitable.

Type (ii) and (iii) are suitable for general masonry work in winter. They have lower strength than type (i) but the lime acts as a plasticiser in the fresh

Table 3.2    BS 5628, Part 3, mortar types and chief properties.

| | | Mortar designation | Type of mortar (proportions by volume) | | |
|---|---|---|---|---|---|
| | | | | Air-entrained mixes | |
| | | | Cement:lime: sand | Masonry: cement:sand | Cement:sand with plasticiser |
| Increasing | ▲ Increasing ability | (i) | 1:0 to ¼:3 | | |
| strength | to accommodate | (ii) | 1:½:4 to 4½ | 1:2½ to 3½ | 1:3 to 4 |
| and | movements due | (iii) | 1:1:5 to 6 | 1:4 to 5 | 1:5 to 6 |
| improving | to temperature | (iv) | 1:2:8 to 9 | 1:5½ to 6½ | 1:7 to 8 |
| durability | and moisture ▼ | (v) | 1:3:10 to 12 | 1:6½ to 7 | 1:8 |

Direction of change in properties is shown by arrows

Increasing resistance to frost attack during construction →

← Improvement in adhesion and consequent resistance to rain

*Note:* The range of sand contents is to allow for the effects of the differences in grading upon the properties of the mortar. In general, the lower proportion of sand applies to grade G of BS 1200 while the higher proportion applies to grade S of BS 1200.

material and helps bond the units and resist water penetration in the hardened material. (Lime is in fact hydrated lime, $Ca(OH)_2$, normally sold as a bagged powder which eventually contributes to strength by carbonation (see equation on p. 69). Most limes are non-hydraulic – they do not set by addition of water – though some limes containing impurities such as magnesium carbonate, as obtained in dolomitic limestone, do have some hydraulic action.)

Types (iii), (iv) and (v) mortars contain progressively more lime to compensate for the reduced cement content in plasticising terms. Type (iv) mortars can be used for general purposes in summer while type (v) are restricted to internal use because they have low durability. Note that when bonding weak products, such as lightweight blocks externally, low-strength mortars are preferred as protection would normally be provided by a cladding or rendering. Where low-strength stock facing bricks are used, a higher strength pointing can be used for protection, though such pointing is not recommended in highly stressed situations; it tends to transmit stress through the brick surfaces, being stiffer than the remaining material, and may lead to spalling of the brick surface.

Instead of lime, liquid plasticisers may be used (Table 3.2), having an air-entraining effect and, therefore, improving frost resistance, though mortars so produced are less cohesive in the fresh state and less impermeable in the hardened state. Recent research has shown that excellent 'all-round' performance is obtained by incorporating a liquid plasticiser in a $1:1:5\frac{1}{2}$ cement/lime/sand mortar – leading to the advantages of both lime and liquid plasticisers. A further alternative is supplied by masonry cements which have properties intermediate between the above types (Table 3.2). Lime mortars are now less widely used than previously since lime is bulky and unpleasant to handle, though in some situations, for example, in exposed conditions with low-suction bricks, their use is recommended to help avoid water penetration.

As well as the functional requirements described above, mortars should contribute to the aesthetic quality of the walling. The commonest problem is associated with colour; to achieve the best appearance, uniformity of colour from batch to batch is essential. This requires consistency of colour of sand, cement and, if used, pigment, together with accurate batching. Sand and cement should be obtained in adequate quantity to ensure uniformity during construction. Sand quality is often variable and although unwashed 'soft' (building) sands are generally suitable, they should be checked for excessive clay content or fine material (BS 1200). To achieve repeatable mixes, the use of gauging boxes for batching is recommended. Machines are now available for automatic batching and mixing of mortars and these should give successive batches of very uniform quality and colour. A further option is the use of ready-to-use mortars in which case the supplier should guarantee consistency of colour. These mortars are set-retarded so that they can be used up over a working day.

There is concern that, with steadily increasing demands for good thermal insulation, the mortar joints cause quite extensive cold bridging, especially in solid wall construction, increasing heat transmission by as much as 25 per cent. This can be overcome by the use of recently introduced lightweight mortars in which sand is replaced by a suitable crushed lightweight aggregate, such as perlite. It should be possible, by careful mix design, to satisfy the above functional requirements with the much lower heat transmission which is characteristic of lightweight materials.

As a final cautionary note, it should be appreciated that mortars exhibit very high shrinkage, often over $2000 \times 10^{-6}$. This is of little consequence when correctly used in small thicknesses (up to 15 mm) since a network of fine cracks results. Mortars should not, however, be used to fill large spaces, especially in one operation; concrete or other materials with relatively low shrinkage would perform such functions much more satisfactorily.

## Renderings

These may be defined as cement-based mortar coatings for external or other surfaces which must be water-resistant. They have the following chief functions.

* To provide an aesthetically pleasing, easily maintained surface finish.
* To resist rain or damp penetration.

To achieve these functions, rendering must adhere well to the background and be reasonably impermeable and free of cracks.

The mortars used for rendering are basically the same as those for brickwork, though careful attention should be paid to sands (see BS 1199). Coarse sands will lead to mortars with poor adhesion while very fine sands or sands containing clay require more water and lead to shrinkage problems. As with jointing mortars, the mortar should not be stronger than the background. Lime is again considered a valuable material, especially in weaker mortars.

The correct technique for rendering involves the application of at least two coats. The first coat is designed to level the background and even out its suction or water absorption, which may vary considerably from place to place, being lower in the case of lintels, mortar joints and dense bricks, and higher with lightweight blocks. Treatment of the last with PVA emulsion may be necessary to reduce suction to acceptable levels. Saturation with water is not recommended, as it leads to later shrinkage. Raking of mortar joints is essential for good adhesion if the render is to be applied to a smooth background such as common bricks. The 'ideal' thickness for the undercoat is about 15 mm.

Adequate trowelling in the undercoat is very important; this compacts the mortar, closing cracks as the water is drawn into the background. Trowelling would normally be completed within about one hour of application. The undercoat is then 'scratched' to assist in producing a uniform distribution of small cracks and to provide a key for the final coat. Curing for at least three days is necessary; a weak and friable product results if there is inadequate water for cement hydration. Premature drying is easy to spot since the material becomes much lighter in colour as it dries. After curing, drying should be allowed in order to permit a network of fine shrinkage cracks to form.

The final coat should be applied after the first coat has cured and dried, usually in about one week, and may take several forms:

- *Smooth rendering.* A smooth finish is obtained by trowelling, with a wooden float to about 3 mm thickness. Metal trowelling produces a very smooth finish that is more likely to be affected by crazing problems. Smooth rendering can be shaped and ruled to produce the appearance of natural stone at a fraction of the cost; this is known as 'stucco' (Fig. 3.9)
- *Pebble dashing.* Pebbles of size 3–10 mm are dashed into the final coating while plastic.
- *Rough cast.* Mortar is dashed against the wall producing a textured finish. The coarseness of texture increases with the maximum aggregate size.

Fig. 3.9  Stucco finish to buildings in Brighton. Lower elevations are ruled to produce an ashlar (coursed stone) appearance. Scrolls and other decorative details are moulded.

Some cracking invariably occurs in the final coat but is obscured in coarse-textured finishes. In smooth-rendered finishes, a very fine crack network results which should be invisible if the final coating is only a few millimetres thick. Smooth renderings are normally finally decorated with a masonry paint which will fill and obscure such fine cracks. Curing of the final coat is essential in all cases for best results.

The production of good-quality renderings requires skill and experience as well as organisation of work, for example, the final coat should be applied in one continuous operation to each elevation in order to avoid unsightly joins in the material.

## Experiments

## Experiment 3.1   Determination of the compressive strength of clay bricks

The experiment is much more valuable if 10 bricks of a given type are tested, since the variability of the bricks can then also be ascertained. If this is not possible, a smaller number of, say, three bricks might be tested to give some indication of the ranges.

*Apparatus*
Gauging trowel, four 75 mm cube moulds for mortar, curing tank; compression tester; plywood sheets of size 225 × 112 mm and thickness 4 mm – two per brick tested, assuming standard metric bricks.

*Preparation of bricks*
All bricks should first be soaked for 24 h and then removed and allowed to drain for 5 min.

If bricks have frogs which are to be laid frog-up, the frogs must be filled with mortar prior to testing.

The mortar should comprise a 1 : 1½ cement : sand mix with sufficient water for satisfactory workability. (To achieve the required 3 to 7-day strength with very fine sands, it may be necessary to use a 1 : 1 cement : sand mix). Fill the frogs with mortar and, after allowing 2 to 4 h for stiffening, trowel as smooth and flat as possible. Make four 75 mm cubes from the mortar. Store the bricks and cubes under damp sacking for 24 h and then demould cubes and transfer when the mortar strength is between 28 and 42 N/mm$^2$ (usually achieved in 3 to 7 days).

*Procedure*
Remove bricks from the water, wiping away excess water and any grit. Label the bricks and record the dimensions of the smaller bed face of each brick. Mount centrally in the compression machine with plywood sheets above and

below. Bricks should be mounted in the same orientation as they are intended to be used in service (normally frog-up). Apply the load at 15 N/mm$^2$ per minute (approximately 33 kN per minute for standard bricks) until maximum load is reached. Record this load and calculate the stress at failure, using the dimensions of the smaller bed face. Repeat for the other bricks and find:

* the average strength
* the range of strength

Hence classify the brick (see page 94).

## Experiment 3.2   Determination of the water absorption of clay bricks

To obtain a meaningful result, the BS 3921 boiling or vacuum test should be carried out. The vacuum test is much quicker than the boiling test but requires additional apparatus so that the latter test will be described here for simplicity. It is suggested that two brick types should be investigated – for example, Flettons and Engineering bricks.

*Apparatus*
Weighing balance capable of 0.1 g accuracy; tank of capacity (ideally) sufficient to heat 10 specimens; drying oven.

*Procedure*
Ten bricks should, ideally, be used, though they may be sawn into half or quarter sections to save space in the heating tank.
   Dry the bricks for 2 days at 100 °C and then allow to cool to room temperature. Label and weigh the bricks and record the masses. Place in the water tank so that water can circulate freely. Heat to boiling and maintain the temperature of 100 °C for 5 h. Allow to cool *still submerged in the tank* for 16–19 h. Remove the bricks, wipe off surface moisture and weigh immediately.
   Calculate the water absorption (see page 92) and compare with the limits for Engineering bricks (Table 3.1)

## Experiment 3.3   Measurement of efflorescence in clay bricks

In common with other tests, 10 specimens should be used, since results often vary considerably from brick to brick.

*Apparatus*
Ten wide-mouthed bottles or flasks; large flat tray.

*Procedure*
The test should be carried out in a warm, well-ventilated room. Wrap the bricks in polyethylene sheet, leaving only the face that is to be exposed in the

brickwork uncovered. Tie with a rubber band or string (see Fig. 3.4). Fill the first bottle with distilled water and place a brick over the mouth so that it completely covers the opening. Invert both brick and bottle while maintaining them in close contact and stand in the tray. Repeat for the other bricks. Leave for a few days, replacing the distilled water if it is completely absorbed within 24 hours. When the bricks have dried, add more water and repeat the process. After drying for the second time, examine the bricks for efflorescence and classify, using the criteria given on page 92.

## Experiment 3.4   Measurement of frost damage to clay bricks

Frost usually takes a number of years to result in visible damage to brickwork in practice, so that a laboratory test must attempt to accelerate the process. A fully realistic test must include the effect of mortar joints also – panels comprising 30 or more bricks would be ideal but would require a very large refrigerated enclosure. This experiment is designed to give some indication of the likely properties of a brick, using only a small domestic freezer, though a period of some weeks will still be necessary for its execution.

*Apparatus*
Shallow containers, domestic freezer (upright type is more convenient), compression test machine.

*Procedure*
Ideally 10 or more bricks of each type should be subjected to freeze/thaw cycles, otherwise variability between individual bricks may overshadow the effects of frost, especially in naturally durable bricks.

Take a batch of 20 bricks of the type required and place half under water at room temperature for the duration of the test. Place the other 10 in shallow trays in the freezer. Subject these bricks to temperature cycles in the range 20 to $-20\,°C$, keeping continuously saturated. With an upright freezer one cycle per day should be possible – for example, switch on at 17.00 hours and by 09.00 hours the following day the temperature of $-20\,°C$ should have been reached. At this point, switch off the freezer and leave the door open so that the bricks thaw *fully* – a period of 8 h should suffice. (The thawing cycle may have to be lengthened for chest freezers, which are slower to thaw unless the bricks are removed.) Repeat the procedure, topping up trays with water as necessary.

Inspect every 10 cycles for signs of visible damage. After 50 cycles, thaw the bricks and test in a compression machine (see Experiment 3.1). Measure the average strength of the bricks which were subject to freezing and express as a percentage of that of the remaining bricks.

This is a severe test and, although no clear criterion exists for frost-resistance, bricks might be assumed to have a high degree of frost-resistance if their strength after 50 cycles has not decreased by more than 10 per cent and there is no visible flaking.

## Experiment 3.5    Measurement of crystallisation damage in clay bricks

Crystallisation damage, in common with frost damage, usually occurs progressively over a number of years and this simple test is therefore designed to accelerate the process, though it may still take several weeks to complete.

*Apparatus*
A large tray in which 10 or more bricks can be immersed; plastic rack or grid on which bricks can be dried; compression machine; 5 per cent solution of magnesium sulphate.

*Procedure*
To accelerate drying, the test should be carried out in a warm, well-ventilated room. Place 10 bricks in the solution so that they are completely covered. Place another 10 bricks of the same type under water for the duration of the test. After a few hours, remove the bricks from the solution and place on a rack or stand which allows them to drain freely. Allow to dry completely – this should be easy to ascertain, since dry bricks are usually a lighter colour. The process may take several days, depending on the type of brick and drying conditions.

Repeat the procedure until at least 20 wetting and drying cycles have been applied, topping up the salt solution as necessary.

Finally, rinse in ordinary water and observe any visible effects on the bricks. Test the 10 bricks in a compression machine, as described in Experiment 3.1. Test also the 'control' bricks and express the average strength of the bricks subject to the salt solution as a percentage of that of the others.

Crystallisation may be assumed to have had a significant effect on the bricks if there is visible damage or if the strength has decreased by more than 10 per cent.

### Questions

1.  Describe the production of:

    (a)  Fletton bricks
    (b)  engineering bricks

    Give characteristic properties of both types of brick and two common applications of each.
2.  Explain the meaning of the term *sintering* as applied to clay bricks. Indicate the effects on performance when bricks are:

(a) overfired
(b) underfired

3. Outline the essential requirements of type 'L' clay brick. Give typical applications of these bricks.

4. A clay brick of length 220 mm and width 106 mm failed in compression at a load of 1830 kN. After 5 h boiling in water, the mass increased from 2.94 to 3.05 kg. Use Table 3.1 to classify the brick, assuming it was typical of its type.

5. Explain what is meant by *crystallisation damage* in clay brickwork. Give situations in which it is especially likely to occur. Give three steps which may be taken to minimise the risk of damage.

6. State the cause of sulphate attack in brickwork. Give three precautions by which the likelihood of such attack can be reduced.

7. State what is meant by *efflorescence* and indicate when and where it is most likely to be seen. Explain why efflorescence is unlikely in sand–lime brickwork.

8. Give the main reasons for the current widespread use of autoclaved aerated concrete blocks. Indicate why it is important that these blocks are not saturated prior to use.

9. Indicate four performance requirements of mortars for masonry. State, with reasons, proportions for a mortar mix which would be suitable for

   (a) a wall for a domestic property to be built using 'stock' bricks in winter weather
   (b) an internal wall using autoclaved aerated concrete blocks

## References

### British Standards

BS 187: 1978, *Calcium silicate (sand lime and flint lime bricks)*.
BS 1199 and 1200: 1976, *Specification for building sands from natural sources*.
BS 3921: 1985, *Clay bricks and blocks*.
BS 6073: 1981, *Precast concrete masonry units*.

### Building Research Establishment

BRE Digest 362: 1991, *Building mortar*.

# Chapter 4

# Dampness in buildings

A great many of the problems which commonly occur in buildings are associated with the penetration of moisture. Specific problems will be dealt with in more detail later but it may be worth while to indicate the range of difficulties which could arise. They include fungal attack, swelling/distortion, loss of strength, frost damage, sulphate attack, crystallisation, efflorescence, corrosion of metals and damage to finishes. It is for this reason that great care needs to be given to design, detailing and construction; those buildings which shed rain water (including wind-blown rain) effectively and prevent admission of ground moisture will be largely resistant to the problems listed.

The extent to which water is admitted and retained by the building fabric depends not only on the amount of water that is incident from the exterior of the building (although obviously good design will minimise this), but also on the surface characteristics of the materials used. The most common means of admittance of water is by capillarity.

## Capillarity

In order to explain this term, it will be helpful to consider first the term 'surface tension'.

Surface tension is exhibited by virtually all materials as a result of cohesive forces which exist within them. These are intermolecular forces which arise naturally, due to the electronic nature of the outer shells of atoms. In a liquid, the forces are relatively small and hence they only exert a slight restriction on mobility. Liquids falling freely through space nevertheless take up a spherical shape, since molecules at the surface of the liquid experience a net inwards attraction (Fig. 4.1). The surface is said to be in 'tension' because the inwards attraction results in the body of liquid trying to minimise its surface area, hence producing a sphere – the same shape that would be produced if the surface of the liquid were covered by a thin elastic membrane.

Solids also exhibit surface tension – in fact the cohesive strength of solids can be likened to a sort of surface tension but without the flexibility which exists in liquids. In gases, surface tension effects are very small.

At an interface between a liquid and solid, liquid molecules will be under two influences.

1. Cohesion within the liquid tending to lead to surface tension.
2. Adhesion between the liquid molecules and the solid, caused by the same type of electronic attraction which results in surface tension.

The behaviour of the liquid will depend on the relative magnitude of these two effects. If cohesion within the liquid is greater, it will still tend to minimise its surface area. This occurs when, for example, mercury contacts most solids or when water rests on a greasy surface (Fig. 4.2). When adhesion between the liquid and solid is greater than cohesion within the

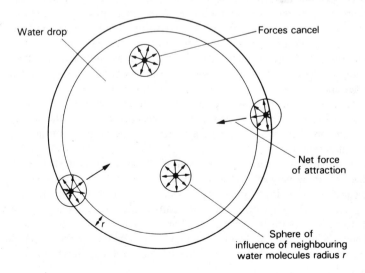

Fig. 4.1  Molecular cohesion – the origin of surface tension.

114

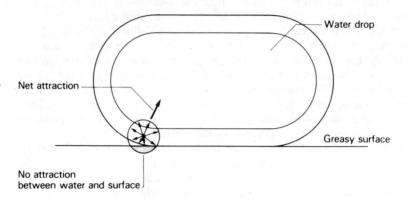

Fig. 4.2   Non-wetting properties resulting when cohesive forces exceed adhesive forces.

liquid, the liquid spreads out on ('wets') the surface, forming only a very thin layer. Reducing the surface tension (cohesion) of water – for example, by adding detergents – increases the relative magnitude of the attraction. Water wets most building materials, including ceramics (such as brick, stone and concrete), wood, metals and some plastics. This attraction is described as *capillarity* and it can theoretically occur in all these materials.

An important factor affecting capillarity in practice is whether there are small spaces or pores associated with individual materials. The effect may be explained by consideration of the water rise up capillary tubes (hence the term 'capillarity') of various diameters (Fig. 4.3(a), Experiment 4.1). It is well known that the capillary rise in glass tubes increases as the tube diameter decreases (Fig. 4.3(b)) because the force exerted on the liquid at the interface increases in relation to its cross-sectional area as the tube diameter decreases. It is true that the total capillarity force involved reduces as the tube gets narrower, since the length of meniscus decreases but the area of the liquid supported decreases at a greater rate, so the result is an apparently higher attraction.

In summarising, it may be stated that, if water wets a solid surface, it will be drawn into any gaps or pores in that material, the attractive force increasing as the width of gaps or pores decreases. In some materials, water can in fact be drawn with some force into cracks or pores which are too small to be seen with the naked eye. For the same reason, such water will also be slow to evaporate from those materials once the source of dampness is removed, since it is held in position by the same capillary forces which caused admission initially.

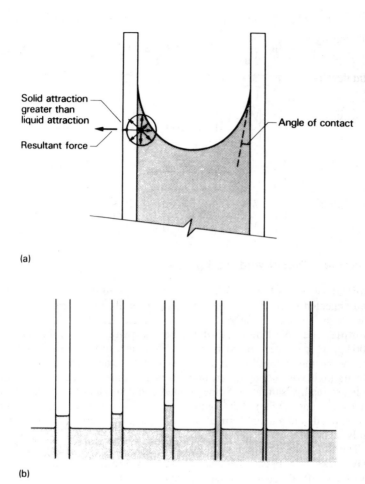

(a)

(b)

Fig. 4.3 Capillarity: (a) attraction of water to the solid surface – the greater the attraction, the smaller is the angle of contact; (b) effect of tube size on capillary rise.

## Porosity

Porosity can be defined as

$$\frac{\text{Volume of pores in a given sample}}{\text{Bulk volume of sample}} \times 100$$

or

$$\frac{\text{Bulk volume of sample} - \text{Solid volume of sample}}{\text{Bulk volume of sample}} \times 100$$

Bulk density ($D_B$) and solid density ($D_S$) can be related as follows:

Bulk density $\qquad D_B = \dfrac{M}{V_B} \qquad M = \text{mass}; V_B = \text{bulk volume}$

Solid density $\qquad D_S = \dfrac{M}{V_S} \qquad V_S = \text{solid volume}$

Porosity $\qquad P = \left[ \dfrac{V_B - V_S}{V_B} \right] \times 100 = \left[ \dfrac{1 - V_S}{V_B} \right] \times 100$

Hence $\qquad \dfrac{V_S}{V_B} = 1 - \dfrac{P}{100}$

Substituting $\qquad D_B = \dfrac{M}{V_B} = \dfrac{D_S V_S}{V_B} = D_S \left[ 1 - \dfrac{P}{100} \right]$

## Interconnection of voids and pore structure

Capillarity can clearly only take place in materials which have interconnected voids. Some voids in most materials are not easily accessible to water and therefore cannot be easily saturated. Taking timber by way of example, the solid density of timber is quite high (approximately $1500 \, \text{kg/m}^3$); if voids filled readily with water, timber would sink quite quickly on immersion. In fact, even low-density wood such as balsa wood, with its high void content, can float for some time, indicating that not all voids are readily saturated. Some aerated concrete blocks will also float on water for a time, again indicating that voids are not easily filled. The voids in air-entrained concrete are formed within the cement mortar and are therefore highly resistant to water penetration.

The importance of pore size distribution has also been mentioned under 'Clay bricks' – it appears that coarse pores (say, greater than 30 nm diameter; $1 \, \text{nm} = 10^{-9} \, \text{m}$), although they may be readily saturated, do not constitute a frost hazard because, on freezing, water tends to flow out of them to surface regions. The main problem occurs with smaller pores from which water tends to flow at reducing temperature, to 'feed' growing ice crystals.

At the present time, no simple way has been found of relating weathering resistance to pore structure, and consequently materials specifications for durability must relate to other aspects of materials properties which control durability rather than to terms such as porosity. It was stated earlier that, in the case of clay bricks for example, some types of brick which are highly absorbent nevertheless satisfy the frost resistance requirement of BS 3921. In some cases, an absorbent brick can indeed increase the durability of the wall it is part of – for example, the resultant 'suction' of the brick can increase the cement mortar bond and thereby reduce the likelihood of water penetration at the brick/mortar interface. An absorbent brick will also tend to reduce the moisture content of the mortar joint – often the most vulnerable part of the wall (see Experiment 4.3).

A related property of materials is *permeability*, which is a measure of their ability to transmit water vapour or other substances under pressure. The term is very relevant in sheet materials such as vapour barriers or coatings such as paints. The term *vapour resistance* is also used. This is defined as:

$$\text{Vapour resistance} = \frac{\text{Thickness}}{\text{Permeability}}$$

and is therefore applied to films of specified thickness. As an example, the vapour resistances of various films/coatings to water vapour are:

| *Film* | *Vapour resistance* (GNs/kg) |
|---|---|
| Microporous paints | 1–4 |
| Gloss paints | 5–8 |
| 0.15 mm polythene film | 350 |
| Aluminium foil | 4000 |

Water admission can be reduced by such films in appropriate situations. Table 4.1 summarises some of the problems which may arise in materials into which moisture has penetrated.

## The prevention of moisture penetration in buildings

The following are intended to illustrate the various mechanisms by which moisture may penetrate buildings.

### Cavity walls

There should be no path for moisture between the outer and inner skins. Hence cavities must be kept free of mortar droppings, vertical dpc's should be provided in reveals (Fig. 4.4) and wall ties should be designed to shed water (Fig. 4.5). Complete bridging of cavities with plastic foams or other materials for thermal insulation purposes will increase the risk of moisture penetration in exposed areas (see BS 5618). Preformed types which leave part of the cavity unfilled if used with special wall ties are to be preferred.

### Brickwork

Moisture can penetrate by capillarity into the mortar and the brick individually, but also at the brick/mortar interface unless there is good bonding at this position. To reduce moisture penetration into brickwork:

• Provide adequate eaves projection to shed water clear of the brickwork.

Table 4.1 Effects of water admission in porous materials

| Effect of dampness | Materials involved | Result |
|---|---|---|
| Fungus attack | Timber | Softening, eventual failure |
| Swelling/ distortion | Timber | Distortion, poor fit of windows, doors etc. |
| | Brickwork | May cause buckling if inadequate movement joints. |
| Strength loss | Brickwork, concrete | Not normally a direct cause of failure |
| | Timber | May cause premature failure or increase deflections |
| Frost damage | Brickwork, concrete, mortar | Spalling followed by crumbling and failure at later stage |
| | Clay tiles | Delamination, eventual failure |
| Sulphate attack | Concrete, mortars | Disruption followed by failure |
| Crystallisation | Clay brickwork | Disruption followed by failure |
| Efflorescence | Brickwork, plaster and concrete | Unsightly appearance |
| Corrosion of embedded metals | Mainly steel (aluminium if in contact with cement). Examples – wall ties, steel reinforcement in poor quality concrete, screws, nails, etc. | Impaired efficiency followed by failure |
| Damage to finishes | Wallpaper, emulsion paints, fabrics | Discoloration, mould growth, loss of adhesion |
| | Oil paints on cement substrate | Saponification (alkali attack) |

- Avoid recessed pointing which would provide horizontal surfaces to receive water.
- Ensure adequate bond between bricks and mortar (the use of lime as the mortar plasticiser may help).
- Use damp-proof courses correctly positioned at low level and at high level for parapet walls (Fig. 4.6).

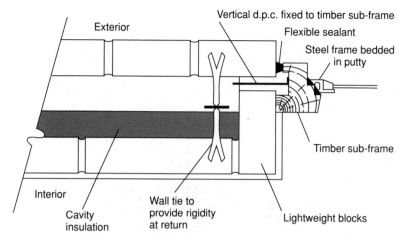

Exterior

Vertical d.p.c. fixed to timber sub-frame

Flexible sealant

Steel frame bedded in putty

Timber sub-frame

Interior

Cavity insulation

Wall tie to provide rigidity at return

Lightweight blocks

Fig. 4.4    Vertical dpc to prevent water penetration at window return.

Fig. 4.5    Wall ties to prevent passage of water across cavity.

- Use weathered (sloping) coping where tops of walls are exposed (Fig. 4.6).

## Windows and doors

Rotting of window and door frames is one of the most common problems in building. The likelihood of this can be reduced by:

- Use of glued joints in joinery to prevent water access when paint cracks.
- Use of anti-capillarity grooves (Fig. 4.7).
- Recessing frames from wall surface.
- Use of sealant around frame edges.
- Inward opening of external doors – these are then less vulnerable.
- Provision of weather bar to shed water clear of door bottoms.

## Roofs

The pitch of a roof must be related to the roof finish, the object always being to minimise water penetration into the surface material. Taking plain tiles

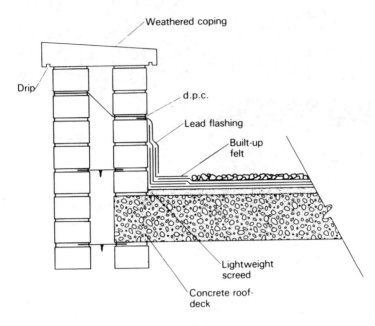

Fig. 4.6   Weatherproofing details in parapet wall (additional insulation may be required).

(Fig. 4.8) as an example, important features are:

- A minimum pitch, usually of 35°, to reduce water absorption.
- Tiles are curved to reduce capillarity between them.
- No joints above one another (requires 'one and a half' width tiles at each end of alternate course).
- Fascia board upstand to prevent capillarity between eaves course tile and first full tile course.
- Roofing felt under tiles to shed wind-blown rain or snow.
- Adequate ventilation of the roof space and eaves.

## Remedial measures when damp has affected masonry walls

Before measures are taken to rectify the effects of dampness, it is essential to identify the cause of the problem. This should be done as soon as possible after dampness is detected since there are often salts present in masonry, soil or adjacent materials, which tend to migrate to points of drying around any sources of dampness. If investigation is not carried out promptly, salts can

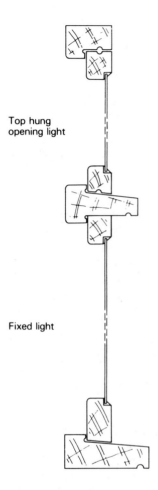

Top hung
opening light

Fixed light

Fig. 4.7   Section through timber casement window showing anti-capillary grooves and drips on weathered transom and sill.

build up to a level whereby they will seriously impair drying even after the fault has been rectified. The first signs of dampness are, quite often, deterioration of internal finishes, usually associated with damage to plaster resulting from migration of salts. Possible sources may be as follows:

• *Leaking rainwater system.* Since very large quantities of water may be discharged from a roof the importance of maintaining gutter and downpipe systems in good repair cannot be overstated. A simple problem such as a blockage due to leaves can result in large quantities of water

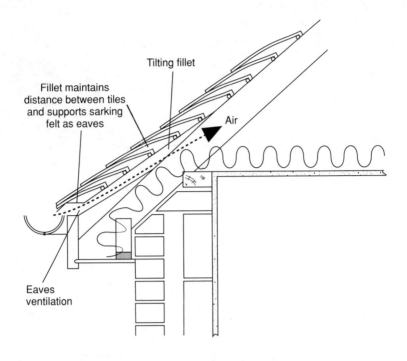

Fillet maintains distance between tiles and supports sarking felt as eaves

Tilting fillet

Air

Eaves ventilation

Fig. 4.8   Detail of plain tiling on pitched roof. Note that the sarking felt is fully supported.

being discharged down a wall face. Only a wall of very high quality would be unaffected in such situations.

- *Rising damp.* This is common in older properties which may not have an effective damp-proof course, or in newer properties in which the dpc has deteriorated. A further common cause is too high a ground level, or bridging of the dpc by, for example, steps.
- *Defective damp-proofing properties of the wall.* Many older properties do not have cavities, relying on the 'overcoat' effect, whereby water is absobed by a porous wall material in wet weather, the water drying out in dry weather spells. Alternatively, and quite commonly, many properties have defective cavities, usually resulting from poor workmanship during construction as a result of which cavities become bridged. A further possible cause is inappropriate use of cavity fills; urea formaldehyde foams, for example should not be used in situations of high exposure.
- *Leaking chimneys.* Chimneys are highly exposed structures so that effective damp-proofing measures must be taken during construction. These include the use of strong sulphate-resistant mortars to resist penetration of water at mortar joints, together with suitable

flashings/soakers. Poorly applied renderings can trap water, rather than resist its ingress.

- *Condensation.* Since water vapour levels are generally higher inside dwellings, water has a tendency to diffuse outwards through the walls. 'Cold' internal wall surfaces are very easily created by, for example, positioning wardrobes on outside walls, water vapour then condensing on the wall, behind the wardrobe. This situation will be exacerbated if vapour levels in the property are already high; large numbers of house plants are one common cause of high vapour levels.

Remedial treatment for many of the above problems will be self-evident, though some of the problems merit further comment.

Rising damp can be quite difficult and expensive to rectify if the origin is lack of a suitable dpc. A number of systems for rectification are available, including injection of silicones or chemicals to create a barrier to moisture; the use of ceramic tubes which draw water towards them, subsequently allowing it to evaporate; and 'electrical' methods. It is advisable in every case to ensure that techniques have appropriate certification, and that they are carried out by a competent body, with guarantees.

Localised repairs to wall cavities containing obstructions are possible though the appearance of the masonry unit may be affected by such repairs. The repair of larger areas is likely to be more difficult; use of water-repellant coatings has been tried, but it is difficult to prevent water penetration in brickwork by such techniques because most brickwork contains small cracks which readily admit water by capillarity even after treatment. A possible solution to rain penetration would be rendering (to which, if there is a history of problems, a waterproofing admixture should be added). Where renderings are defective they should be replaced.

Where internal plaster has been affected it should be removed. This will also remove any salts which often accumulate in the plaster. The background can then be primed with a moisture-resisting form of PVA (polyvinyl acetate) emulsion to prevent any residual dampness from affecting the new plaster. The wall can then be rendered with a sand/cement mortar, containing a waterproofer, a class C plaster being used for the 'set' (final) coat – see Chapter 10. Alternatively, gypsum-based 'renovation' plasters are available; these are designed to dry reasonably quickly in spite of there being some residual moisture in the wall. Whatever system is employed, it is important to deal with the problem at source – that is, to overcome the original cause of the dampness.

## Experiments

### Experiment 4.1   Determination of the effect of pore diameter on capillary attraction and the magnitude of the surface tension of water

*Apparatus*
Approx. 200 mm lengths of capillary tubes in a range of bore diameters, for example 0.2, 0.5, 1, 2 mm; beaker; clamp stand; acetone; distilled water; travelling microscope or accurate weighing balance.

*Procedure*
Clean the capillary tubes thoroughly with acetone, rinse with distilled water and dry. Do not touch either end of the tube once cleaned. Place each tube in turn in a beaker of distilled water filled to the brim and fixed in a vertical position with a clamp stand. Measure the capillary rise to the nearest mm. Repeat for the other tubes. The capillary rise should increase progressively as the tube diameter decreases – if it does not, reclean the tubes and repeat.

Measure tube diameter with a travelling microscope, or by accurately weighing the tube dry and then with the bore filled with water.

$$\text{Diam (mm)} = 1000 \sqrt{\left[ \frac{\text{Mass increase (g)} \times 4}{\pi \times \text{Length of bore (mm)}} \right]}$$

(Alternatively, bore diameters may be given.)

Plot the capillary rise graphically ($y$ axis) against bore diameter ($x$ axis). Notice the trend in the capillary rise for very small bore diameters – pores in building materials may be as small as 10 nm (= 0.01 µm = $10^{-8}$ m). The surface tension ($T$) of water can be calculated from the formula:

$$T = \frac{R\rho gh}{2}$$

$R$ = radius of bore

$\rho$ = density of water (1000 kg $(m^3)$)

$h$ = capillary rise

### Experiment 4.2   Measurement of bulk density, solid density, porosity and degree of saturation of porous building materials

*Apparatus*
Oven; beaker; relative density bottle; balance accurate to 0.1 g and a further balance accurate to 0.0001 g; clamp stand; cling film; fine thread; mortar and pestle; 150 µm sieve; funnel.

*Specimen preparation*
Pieces of brick, concrete or insulation block of size about 50 × 50 × 50 mm are suitable. It is helpful if materials which float on water are cut accurately so that the bulk volume can be determined by dimensional measurement.

*Procedure*
Dry specimens thoroughly in an oven at 110 °C and allow to cool in air.

*To measure bulk density ($D_B$)*
Weigh specimen to 0.1 g accuracy. Find bulk volumes as follows:

(a) Specimens which float in water – for example, aerated concrete blocks: Obtain the volume by dimensional measurement

$$\text{Vol. in litres} = \frac{\text{Vol. in mm}^3}{10^6}$$

b) Specimens which sink in water: Wrap carefully in cling film. Fill a beaker with sufficient water to enable the specimen to be totally immersed in it without causing overflowing. Place this beaker in a top loading balance and note the mass in g (0.1 g accuracy). Wrap the specimen carefully in cling film, making sure there are no trapped air pockets, and attach a thread. Suspend, using a clamp stand, in the beaker of water, avoiding contact with sides or base. Note the new balance reading in g.

Bulk volume in ml = (New reading – Old reading)
Hence, find bulk density:

$$D_B = \frac{\text{Mass (g)}}{\text{Bulk volume } in \ litres} \quad \text{g/l or kg/m}^3$$

*To measure solid density ($D_S$)*
Pulverise about 20 g of the material, using a mortar and pestle, until it passes a 150 µm sieve. Weigh out about 10 g of the material, using the sensitive balance (mass $M_1$ g) and transfer, using a funnel, to the relative density bottle, avoiding spillage. Add one or two drops of soap solution to a beaker of distilled water, stir and then half fill the relative density bottle, using the funnel. Swirl gently until *all* solid material has sunk and entrapped air is removed. Fill to overflowing, insert the stopper and dry the outside of the relative density bottle. Weigh ($M_2$ g).

Empty and clean the bottle, refill with distilled water, insert the stopper, dry and weigh ($M_3$ g).

The solid volume of the material is

$$M_1 - (M_2 - M_3) \ \text{(ml)}$$

The solid density ($D_S$) of the material is

$$\frac{\text{Mass}}{\text{Solid volume}} = \frac{M_1}{M_1 - (M_2 - M_3)} \quad \text{g/ml}$$

(multiply by 1000 to change to kg/m$^3$).

Use the following formula to calculate the porosity ($p$)

$$D_B = D_S \left[ 1 - \frac{p}{100} \right]$$

*To find the water absorption and the percentage pores filled on saturation*

The sample used for the bulk density determination (oven dry mass $W_1$ g) can also be used for this determination. To obtain an indication of the degree of saturation obtained by short exposure (say, to rain), immerse in water for a period of about 10 min. Weigh again ($W_2$ g).

The percentage water absorption $= \dfrac{W_2 - W_1}{W_1} \times 100$

The degree of saturation is found as follows:

Bulk volume of this sample $= \dfrac{W_1}{D_B}$ litres ($D_B$ in g/l)

Solid volume of this sample $= \dfrac{W_1}{D_S}$ litres ($D_S$ in g/l)

Volume of pores equals bulk volume minus solid volume

$$= W_1 \left[ \frac{1}{D_B} - \frac{1}{D_S} \right] \text{ litres}$$

or

$$1000 W_1 \left[ \frac{1}{D_B} - \frac{1}{D_S} \right] \text{ ml}$$

Volume of water absorbed $= (W_2 - W_1)$ ml
Hence, percentage of pores filled (degree of saturation)

$$= \frac{W_2 - W_1}{1000\, W_1 \left[\dfrac{1}{D_B} - \dfrac{1}{D_S}\right]} \times 100$$

This experiment should be repeated, boiling the sample for some time to determine the total proportion of pores that is accessible to water. The performance of different bricks or block types should also be compared.

## Experiment 4.3    Measurement of water absorption of different types of mortar

*Apparatus*

70 mm mortar cubes; gauging trowel; weighing balance; plasticiser; curing tank; oven.

*Procedure*

Make up samples of mortar using the following quantities of cement and builder's sand.

| Ratio cement : sand | 1 : 3 | 1 : 6 | 1 : 9 | 1 : 12 |
|---|---|---|---|---|
| Cement (g) | 250 | 150 | 100 | 80 |
| Sand (g) | 750 | 900 | 900 | 960 |

Mix each into a workable mortar with water which has been plasticised either with a mortar plasticiser or a few drops of washing-up liquid. The consistency of each should be the same.

Fill 70 mm mortar cubes with each, identify and cover.

Demould after one day and cure under water until one week old.

Oven-dry the cubes. Weigh each cube and record the masses. Immerse in water and leave for 5 min. Record masses again. Re-immerse and obtain masses after 10, 20 and 60 min. Calculate the water absorption of each, as a percentage, at each stage.

$$\text{Water absorption} = \frac{\text{Water absorbed}}{\text{Dry mass}} \times 100$$

Plot water absorption against time. Comment on the relation for the different mortar types and give implications regarding the use of these mortars.

## Questions

1. State the meaning of the following terms:

   (a) cohesion
   (b) adhesion

   Hence explain the meaning of the terms *surface tension* and *capillarity*.
   Give simple illustrations of the behaviour of water which results from these two phenomena.

2. Describe how the degree of attraction of water into water pores varies with pore diameter. Hence indicate whether coarse- or fine-pored materials tend to be quicker drying after being saturated.

3. A lightweight block is found to have a solid density of 2400 kg/m$^3$ and a bulk density of 800 kg/m$^3$. Calculate its porosity.

4. A common brick has a bulk density of 1300 kg/m$^3$. If the porosity is 45 per cent, calculate the solid density of the material.

5. Explain the meaning of the term *suction* in bricks. Give two advantages that result from bricks having some suction.

6. Draw a section through a 1 m high, 215 mm free-standing boundary wall, indicating features designed to minimise water admission:

   (a) from above
   (b) from wind-blown rain
   (c) from the ground

7.  (a)  Show by means of a sketch how plain roofing tiles resist capillarity.
    (b)  Give two reasons why tiles on a more steeply-pitched roof are less likely to be damaged by frost.

8.  Give four possible causes of damp patches on internal plaster, indicating areas most likely to be affected by each type.

## Reference

### British Standard

BS 5618: 1985, *Code of practice for thermal insulation of cavity walls (with masonry or concrete inner and outer leaves) by filling with urea formaldehyde (UF) foam systems.*

# Chapter 5

# Metals

## Introduction

In spite of their high cost, there will always be a place for metals in the construction industry because:

- Metals are generally ductile – they exhibit the phenomenon of being able to undergo substantial plastic flow without damage. In this respect they are unique among stronger materials types.
- They have a wide range of properties according to type.
- The stronger metals have higher stiffness and tensile strength than most non-metals.
- They are non-porous, hence chemical activity tends to be confined to the surface.
- Metals are easily alloyed, further increasing versatility.
- Strong bonds can be produced easily by soldering, brazing or welding.

However, some problems are associated with metals; for example, the metallic state is a high-energy state, as a result of which metals tend to

- Be expensive in energy terms to produce.
- Deteriorate by chemical surface action in normal atmospheres (corrosion).
- Have high density – even aluminium, with a relative density of 2.7, is denser than dense concrete (relative density about 2.3).

Table 5.1 gives ultimate tensile strengths (that is, breaking strengths) and elastic moduli of the most commonly used building metals, together with their relative densities. It will be noted that there is reasonable correlation between strength and stiffness, since both are related to metallic bond strength, values for iron being substantially in excess of those for the other metals. Density is not, however, so correlated to the other properties. The following comments apply to the strength values in the table:

- The tensile strength of metals may depend on the cooling regime – more rapid cooling generally produces a finer grain structure and hence higher strength.
- Yield stresses, which determine how large a load can be supported in service, are usually substantially less than (approximately half) ultimate stresses.
- Metals such as zinc and lead may, in the long term, 'creep' to failure at much lower stresses than those indicated.
- The presence of impurities or alloying elements generally has a marked effect on strength. Most impurities increase strength up to a point but they also increase brittleness.

## Steel

Steel, the most important ferrous metal in building construction, is an alloy of iron and carbon. Its widespread application for structural uses is on account of the following:

- Iron is relatively abundant in the earth's crust; second only to aluminium.
- Its melting point is sufficiently high to produce high strength, but not so high that there are problems reaching this temperature in bulk furnaces.
- Iron undergoes a change of crystal structure at red heat which assists in many production processes and leads to metallurgical advantages.

Table 5.1 Properties of chief building metals (commercially pure). Metals are listed in order of increasing ultimate tensile strength

| Metal | Relative density | Elastic modulus ($kN/mm^2$) | Ultimate tensile strength ($N/mm^2$) |
|---|---|---|---|
| Lead | 11.3 | 16.2 | 18 |
| Zinc | 7.1 | 90 | 37 |
| Aluminium | 2.7 | 70.5 | 45 |
| Copper | 8.7 | 130 | 210 |
| Iron | 7.8 | 210 | 540 |

The presence of carbon in iron leads, at ordinary temperatures, to the formation of a non-metallic compound – iron carbide (cementite) – which is intensely hard and brittle, and exists as very fine sheets interwoven with iron sheets to form a compound called pearlite. The advantage of having some pearlite in steel is that plastic flow, or 'yield', occurs much less readily than in pure iron which contains no pearlite, though the presence of too much pearlite reduces ductility, toughness and weldability. The quantity of pearlite contained in a steel depends on its carbon content. The average carbon content of pearlite is 0.83 per cent (by weight), so that steels having this percentage of carbon would consist of 100 per cent pearlite. Steels containing less than 0.83 per cent carbon comprise a mixture of pearlite and iron crystals, the amount of pearlite increasing proportionately as the carbon content rises to 0.83 per cent. Steel may contain up to 1.7 per cent carbon, the carbon in excess of 0.83 per cent in steels of carbon content in the range 0.83 to 1.7 per cent existing in the form of thin cementite layers around pearlite 'crystals'. Figure 5.1 shows magnified photographs of polished, etched sections of steel having various carbon contents.

## Low-carbon steels

These may be defined as steels containing less than 0.25 per cent carbon and include steels referred to as mild steels and high-yield steels. They form a very large proportion of the steels used in building, having moderate strength but good toughness, ductility and welding properties on account of the fairly low proportion of pearlite. Examples include weldable structural steels, cold-rolled steel sections for lintels, etc., window frames, reinforcing bars and mesh, expanded metal and wire for wall ties.

In addition to containing some carbon, low-carbon steels contain a number of other elements, some present as a result of the steel-manufacturing process and others included to benefit certain properties of steel.

Common impurities are:

- *Sulphur and phosphorus.* These form precipitates of iron sulphide (FeS) and iron phosphide ($Fe_3P$) respectively, which cause embrittlement. Hence contents of each are limited to 0.05 per cent in mild steel.
- *Silicon.* When present in small quantities, it dissolves in the iron, increasing strength. Large quantities cause embrittlement, hence silicon is usually restricted to about 0.2 per cent.

Common alloying elements are:

- *Manganese.* This is useful since it increases the yield strength and hardness of low-carbon steels. It also increases hardenability and may combine with sulphur, reducing brittleness caused by the latter. Mild steel contains approximately 0.5 per cent manganese, while high-yield steel contains about 1.5 per cent.

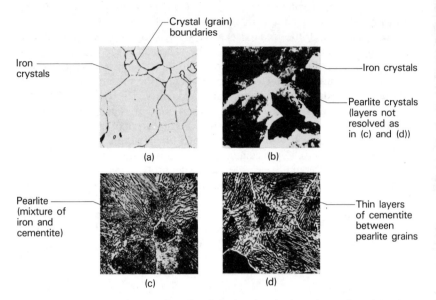

Fig. 5.1  Magnified photographs of polished and etched sections of steel having various carbon contents: (a) carbon content almost zero; (b) carbon content 0.5 per cent; (c) carbon content 0.83 per cent; (d) carbon content 1.2 per cent.

- *Niobium.* This tends to produce smaller crystals (grains), which results in increased yield stress. Quantities in the region of 0.04 per cent may be present.
- *Others.* In addition to the above, small quantities of molybdenum, nickel, chromium and copper may be included in order to obtain specific properties.

The ladle content of a typical high-yield weldable structural steel might be, for example:

$$
\begin{aligned}
C &= 0.21 \text{ per cent} \\
Mn &= 1.5 \text{ per cent} \\
Cr &= 0.025 \text{ per cent} \\
Mo &= 0.015 \text{ per cent} \\
Ni &= 0.04 \text{ per cent} \\
Cu &= 0.04 \text{ per cent}
\end{aligned}
$$

## Heat treatment of steels

One important reason for the versatility of steel arises from the fact that, on heating to red heat (about 900 °C), iron changes its crystalline form from body-centred cubic (BCC) to face-centred cubic (FCC). The latter form has two important differences to BCC iron:

1. It is much more ductile since the crystals contain large numbers of well-defined 'slip planes'. For this reason, large steel sections are always heated to red heat before rolling or other working because at this temperature their working characteristics are more like those of lead. The craft of the blacksmith works on the same principle.
2. It dissolves carbon much more readily – up to about 1.7 per cent by weight.

In order to understand how heat treatment can affect the properties of steels, it is necessary to appreciate that the crystalline form of iron alters each time it is heated or cooled through temperatures in the region of 900 °C. Above this temperature the iron crystals (gamma ($\gamma$) iron) can fully dissolve all the carbon present (up to 1.7 per cent) forming a material called austenite. Hence, each time steel containing up to 1.7 per cent C is heated, the carbon fully dissolves and each time steel is cooled below about 900 °C, low-temperature (alpha ($\alpha$) iron) crystals form as the carbon is rejected to produce a separate compound – iron carbide or cementite – whose physical form has been described. It will be obvious that the formation of new crystals involves the movement of carbon atoms within a solid material and this requires a certain length of time to occur. Hence the properties of the resultant steel can be greatly affected by the rate of cooling. It will also be apparent that steel containing more carbon will be more sensitive to variations in cooling rate since there is more carbon to be rejected (see Experiment 5.1). The following are main forms of heat treatment.

*Annealing*
The steel is heated to about 900 °C so that all carbon dissolves in the high-temperature iron crystals. The steel is then cooled very slowly in the furnace, which results in a coarse crystal structure with relatively thick cementite and iron sheets forming in the pearlite. The result is a soft, malleable steel of fairly low yield point (Fig. 5.2(a)).

*Normalising*
This involves heating the steel as in annealing and then allowing the metal to cool in still air rather than in the furnace. The grains so produced are smaller (Fig. 5.2(b)), as are the cementite layers in the pearlite, resulting in increased strength and impact resistance. Hot rolling of structural steel sections, followed by cooling in air, may give properties similar to those obtained by normalising if the 'finishing' temperature is correct; this then avoids the expense of reheating the steel.

*Quenching*
This involves plunging the red-hot metal into cold water, resulting in a very rapid cooling rate. The carbon in this case is unable to form cementite and so a fine, brittle crystalline structure, called martensite, forms instead (Fig. 5.2(c)). This structure is highly stressed and, in the case of high-carbon

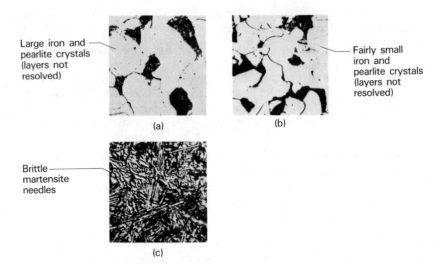

Large iron and pearlite crystals (layers not resolved)

Fairly small iron and pearlite crystals (layers not resolved)

(a)

(b)

Brittle martensite needles

(c)

Fig. 5.2  The effect of cooling rate on a low-carbon steel: (a) annealing – very slow cooling in the furnace; (b) normalising – cooling in still air; (c) quenching – rapid cooling in water.

steels, cracking may occur in the metal. Quenching is, for this reason, followed by tempering – controlled heating to 200–400 °C, to restore toughness to the steel without losing all the hardness imparted by quenching.

It should be emphasised that the effects of rapid cooling are far greater when steels have higher carbon contents, since high-carbon steels have, on cooling, more carbon to 'reject' from their high-temperature crystal structure. Rapid cooling will also be more effective when applied to small components, since they have lower thermal inertia and therefore cool more quickly in given conditions.

## Brittleness in steel

Brittleness implies the inability of the metal to absorb energy by impact (shock loading). Steels which are able to absorb energy in this way would be described as 'tough'. Brittleness depends on:

1. Grain size. It is found that, in the case of ferrous metals, a fine grain size reduces brittleness.
2. Carbon content. Carbon in the form of cementite increases the resistance of the metal to plastic flow at the point of impact. Hence, increased carbon contents tend to increase brittleness, since higher carbon steels contain more cementite.
3. Ambient service temperature. Reducing the temperature increases the yield point of steels, thus making plastic flow more difficult and increasing brittleness.

BS 4360, the British Standard for structural steel, contains a number of toughness categories, the tougher steels being more suitable for service in colder climates or where severe impact is a possibility.

## Ductility of steels

This is a measure of the amount of plastic flow that can be absorbed before failure. It is assessed by measuring the elongation at fracture of a standard length of test piece (Fig. 5.3). Low-carbon steels exhibit highest ductility, especially if annealed since this results in large grains. By carefully controlled cooling, some degree of ductility can also be imparted to high-carbon steels, though these steels are always more brittle than mild steels because they undergo less plastic flow prior to failure. High-carbon steels do not exhibit a visible yield point. Figure 5.4 shows stress–strain curves for mild steel and a high-carbon high-tensile steel, together with curves for copper and lead for comparison.

## Weldability in steels

This is the ability of two pieces of steel to be joined by fusion in a localised region at their boundary. The high temperatures involved in welding may

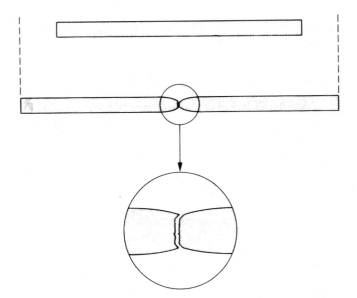

Fig. 5.3  Elongation at failure of a tensile specimen as an indication of ductility. Elongation around the point of fracture is very high in low-carbon steels.

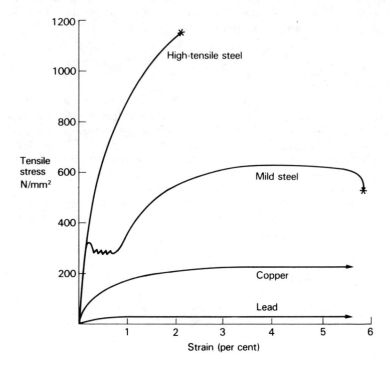

Fig. 5.4 Stress–strain curves for mild steel, high-tensile steel, copper and lead. The last two reach very high strains before failure.

result in fortuitous 'heat treatment' of an area of metal in the neighbourhood of the filler metal – the heat-affected zone (HAZ). Where conditions are such that rapid cooling occurs (mainly in large sections), a brittle martensitic structure results, while slower cooling may cause an undesirably coarse grain structure. Such brittleness problems are most severe in steels having a high carbon or alloy content; steels which are designed to be welded would normally have a specified 'carbon equivalent' content. This is given by the formula:

$$ CE = \frac{\%C + \%Mn}{6} + \frac{\%(Cr + Mo + V)}{5} + \frac{\%(Ni + Cu)}{15} $$

C   =   carbon        V   =   vanadium
Mn  =   manganese     Ni  =   nickel
Cr  =   chromium      Cu  =   copper
Mo  =   molybdenum

The numerical factors applied to the metals other than carbon reflect their decreased tendency to produce brittle martensitic zones.

Few problems should be encountered if the CE is less than 0.25. Higher values of up to, say, 0.5 per cent may be tolerated if cooling is controlled and precautions are taken to keep down the hydrogen content of the weld metal and the HAZ. Hydrogen can be introduced by moisture in some fluxes and it tends to result in cold cracking unless it is dispersed by heat treatment.

## Steel types used in construction

There are very few buildings in which steel is not used in some location as a structural material. The chief applications are as follows:

- in hot-rolled sections in steel framed or other buildings
- as reinforcement or prestressing in concrete framed buildings
- as cavity ties and cold-rolled sections for lintels in masonry construction
- for fixings in timber-framed buildings or timber roofs

The requirements of steel for use in structural sections, the reinforcement of concrete and prestressing of concrete are now briefly considered. For the first two applications, the stiffness ($E$ value) of the steel should be as high as possible, though it is omitted from comments below because it does not vary greatly according to steel type.

### Steel for structural sections (BS 4360)

Such sections are invariably hot rolled, taking advantage of the much greater ductility of steel at red heat. The principal requirements of such sections are:

- *Weldability.* Welding is an economic and extremely effective means of producing structural joints in steel.
- *Ductility/impact resistance.* Steel must be resistant to impact at all service temperatures likely to be encountered and, if overstressed, should yield rather than fail in a brittle manner.
- As far as is consistent with the above requirements, steel should have a high a yield strength.

All BS 4360 steels are of low carbon content, typically 0.2 per cent, to permit welding and produce impact resistance and ductility. A great number of types with differing yield strengths and impact performance are available although grade 43A ('mild') steel is used in almost all normal applications since there are large economic benefits to supplier and user alike in concentrating on a single specification. This steel has a minimum yield stress of around 250 N/mm$^2$ depending on section thickness, with smaller sections tending to have higher yield points. The figure '43' is one-tenth of the minimum *breaking* stress and the 'A' refers to the impact performance. Grade 50A or B steels, which are also available, have yield point stresses of over 300 N/mm$^2$ – the 'B' referring to improved impact performance. High-yield steels may be used where there is a 'gap' in mild steel sizes or where smaller sections are desired at a given loading level. Higher grades can be obtained by adding alloying elements such as manganese or niobium, or by heat treatment such as normalising. Quenching is not feasible in structural-size sections, since the mass of material is too large.

All internal structural steel sections must be adequately protected against fire.

### Reinforcing steels (BS 4449, 4483)

When fixed in place in the formwork, reinforcing steels are *passive*, i.e. there are no applied stresses. Therefore, in order that the steel carries stress in the completed section, loads must be transferred via the concrete. This requires a *bond* between the steel and the concrete. Some of this transfer can be by means of local bond between the steel and concrete surfaces, though in many cases, hooked ends are provided to prevent steel from slipping under stress. The situation is illustrated in Fig. 5.5. With hooked ends but no local bonding the cracking pattern (a) would be obtained in the tension zone, such a large crack tending to admit water, leading to deterioration of steel. Local bonding would result in cracking pattern (b), which is much more acceptable. Note that since the concrete must move in order to transmit stress to the steel, some cracking in the tension zone is inevitable in reinforced concrete. Hence the requirements for reinforcing steels are:

- The steel must bond to the concrete.
- The steel must be ductile to permit bending as well as to guard against brittle failure.
- The steel should be weldable to facilitate joining (though this is rare on site; lapping is normally sufficient). Fabric reinforcement contains welds at each intersection.
- As far as is consistent with the above, the steel should have a high yield strength.

All reinforcing steels are low carbon steels to obtain ductility and weldability. A 'mild' steel is available to BS 4449, and can be recognised by its smooth round section (Fig. 5.6). The advantage of this steel is that tight bends can be formed, a minimum bending radius of twice the bar diameter being possible. This ductility is obtained, however, at the expense of yield strength; the characteristic value is 250 N/mm$^2$. Applications include use in links or where tight bends are required.

High-yield steels are more economic, offering a yield stress of 460 N/mm$^2$ without a proportionate increase of cost. The minimum bending radius is, however, larger at three times bar diameter, as a result of decreased ductility. Three types are available: hot rolled, in which the higher yield strength is obtained by alloying/heat treatment of ribbed bars; and cold-worked bars, obtained by cold twisting/stretching a mild steel bar having

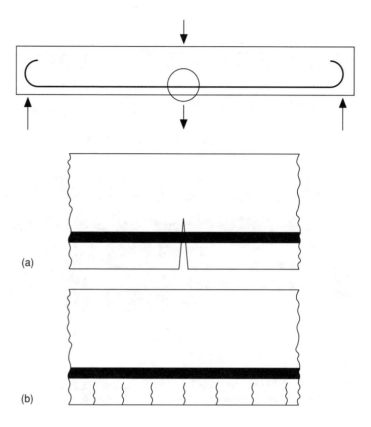

Fig. 5.5 Cracking in tensile zone of reinforced concrete beam: (a) if there is no local bond between steel and concrete; (b) if a local bond between steel and concrete exists.

either a square or ribbed section. The three types of high yield bar are shown in Fig. 5.6. In each case the deformed section provides the additional bond required at higher operating stresses. These types are used for almost all applications.

For applications such as floor slabs, steel fabric can be used. This comprises steel wires welded at the intersections. The welds provide local bond and anchorage.

## Prestressing steels (BS 4486, 5896)

These are so named because stress is applied to the steel before the component is loaded. There are two types: pretensioned and post-tensioned concrete.

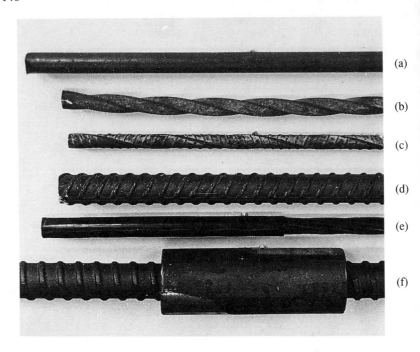

Fig. 5.6 Steels for reinforcing and prestressing concrete: (a) plain round bar (BS 4449); (b) cold worked (type 1) (BS 4449); (c) cold worked (type 2) (BS 4449); (d) hot rolled (type 2) (BS 4449); (e) 7 wire prestressing strand (BS 5896); (f) alloy steel prestressing bar with rolled thread and threaded connector (BS 4486).

In pretensioned concrete the steel is stretched in a rigid frame prior to placing the concrete (Fig. 5.7). When the concrete has hardened the tension is released causing the steel to 'prestress' the concrete. It does so by bonding to the concrete. There are no anchorages at the ends; indeed, the units can be cut to smaller lengths after hardening if required. A problem of all prestressed concretes is that beams contract due to shrinkage and creep, leading to loss of prestress, since the tension in the steel is only maintained as long as the concrete stays at its original length. For this reason it is desirable that the steel tendons are stretched as much as possible at their working load in order that any contraction of the concrete be small compared with the steel extension. Pretensioned components must be made in a factory; the technique is widely used for manufacturing lintels and floor beams. The requirements for steels for pretensioning of concrete are therefore:

- *Extensibility*. The steel must be stretched as much as possible at its working load.
- *Impact resistance*. The steel must be able to withstand vibrations caused by impacts to the stucture;

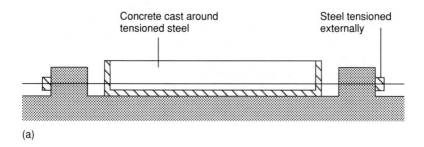

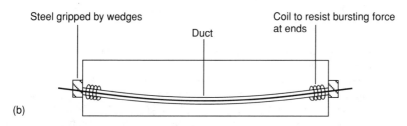

Fig. 5.7    Principles of prestressed concrete: (a) pretensioned; (b) post-tensioned.

* The steel should bond to the concrete.
* As far as is consistent with the above, the yield strength should be as high as possible.

Note that welding and bending are not referred to; the material normally used for pretensioning is wire or strand (Fig. 5.6) which can be coiled and is therefore available in long lengths. There is also no need to bend prestressing steels.

The first requirement – extensibility – would be best fulfilled by use of a low elastic modulus material, but the elastic modulus of steel cannot easily be reduced. However, since there are no welding or bending requirements, the carbon content can be increased to about 0.8 per cent. This, together with heat treatment, gives a high-tensile steel, which, with unchanged $E$ value, also implies a high working strain – over four times that of a mild steel. Typical prestressing wires or strands have a characteristic strength of about 1500 N/mm$^2$, these steels having a 'springy' feel. In the manufacturing process the wire, whether used either as wire, or to form strand, is drawn to increase strength, resulting in a smooth surface. To obtain bond, wire can be indented. Strand has naturally good bonding properties with the concrete.

In post-tensioned concrete beams the tendons are enclosed in a sheath or inserted into a duct in the concrete, not being tensioned until *after* hardening, since the reaction to the steel load is taken by the concrete at its ends. The ends are reinforced to withstand the stresses at each end (Fig. 5.7). The tendons are not normally bonded to the concrete, hence bonding ability is not required. The extensibility arguments used for pretensioning steel also apply. Post-tensioning is widely used for *in-situ* prestressing.

Wire or strand can be used for post-tensioning but for higher load applications high-alloy steel bars can also be used (Fig. 5.6). These cannot be welded but joints can be made since the bars have threads rolled onto them and can be joined by threaded connectors. High-alloy steel prestressing bars of diameter 40 mm have a specified characteristic load of 1300 kN – over five times the load capacity of a similar size mild steel bar.

Prestressed units have the advantage that, unlike reinforced concrete, they can be designed to be completely crack-free, compressive stresses being applied to areas of the concrete, which, under load, would be in tension.

## Oxidation of metals

The strength of the metallic bond arises from the fact that metal 'ions' are completely surrounded by like ions. This cannot occur at the surface of a metal and, in consequence, metals are inherently unstable. In oxygen-containing atmospheres, covalent-type bonds are, therefore, formed between metals and oxygen. Hence metals in the earth's crust are generally found as 'ores' – combinations of metals and other elements of which oxygen is the most common. For example, to extract iron from its ore requires a *reduction* process – the ore is heated with coke, which has a strong affinity for oxygen such that the ore is reduced to the metal, while the coke is oxidised to carbon dioxide.

Once exposed to the air, the oxidation process immediately recommences, although in the case of many metals, including iron or steel at ordinary temperatures, this fortunately occurs at a very slow rate. The restriction on the rate of oxidation of steel results from the difficulty of access of the gas to underlying metal atoms. There follows a progressive reduction in oxidation rate such that, after a time, destruction of the metal ceases. Hence, in a dry atmosphere, steel components will remain sensibly free of corrosion for many years.

The main problem with respect to oxidation occurs at high temperatures at which the oxide film grows much more quickly. Thicker films tend to crack, thereby exposing fresh metal, which continues to oxidise. Hence substantial oxide coatings (mill scale) can form on steel during hot working. This is recognisable as a thin grey coating on hot-rolled steel beams or reinforcing bars. (In cold-worked reinforcing bars, the mill scale falls off during the deformation process.) Similarly, when metals are being joined by fusion, fluxes may be used to prevent oxygen access to the hot metal.

## Corrosion of metals

This process is quite different from oxidation, although oxygen may assist the corrosion process as, for example, in the rusting of iron or steel. Corrosion may take two basic forms – acidic and electrolytic – and these both result from a tendency of metals to ionise (dissolve) when placed in water or aqueous solution (Fig. 5.8). This ionisation is the result of interaction between the surface atoms of the metal (which, as has already been explained, are relatively unstable) and ions in the water. The basic process can be represented as:

| M | $\rightleftharpoons$ | $M^+$ | $+$ | $e^-$ |
|---|---|---|---|---|
| surface | | positive | | electron |
| metal | | metal | | (remains |
| atom | | ion | | on metal) |

The electrons remaining on the metal cause it to become negatively charged, resulting in a negative voltage which increases as further metal dissolves. It eventually reaches a value, called the electrode potential of that metal, at which the negative charge is sufficient to prevent further positive ions leaving the metal, since unlike charges attract. Ionisation (corrosion) ceases at this point.

Continued corrosion, resulting in visible loss of metal, will only take place if, by some mechanism, the negative electric charge on the metal is reduced, thereby allowing more positive ions to 'escape'.

In *acidic* corrosion, the negative charge is removed by hydrogen ions in solution, which then form hydrogen gas:

| 2H+ | $+$ | $2e^-$ | $\rightleftharpoons$ | $H_2\uparrow$ |
|---|---|---|---|---|
| in | | from | | hydrogen |
| solution | | the | | gas |
| | | metal | | |

Metals which, in clean damp conditions, are normally protected by coatings of oxide or other compounds, may nevertheless corrode in the presence of acids such as are obtained in peaty soils (zinc, lead and copper

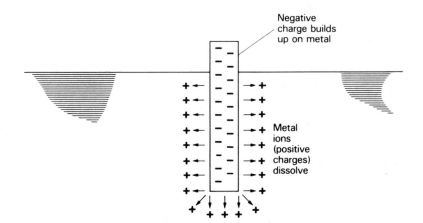

Fig. 5.8   Tendency for metals to dissolve when placed in water.

affected), or when in contact with timbers such as Western red cedar (copper, zinc and lead affected), or rain in polluted environments (zinc affected). Hydrogen is given off *at the point of corrosion* and in severe cases the bubbles may be visible.

*Electrolytic* corrosion occurs when two metals which are effectively different are in electrical contact. This is because the electrode potential varies from one metal type to another. (Table 5.2 gives approximate values for common metals.)

Taking, for example, zinc and copper in electrical contact (Fig. 5.9), the zinc gives a voltage of –0.76 V, while copper gives a voltage of +0.34 V – each with respect to a 'hydrogen electrode'. It may be surprising that there should be positive voltages in the table. This is due to the fact that to measure a voltage two electrodes are needed, and the other electrode will also have some tendency to dissolve. A porous glass electrode containing $H_2$ gas is always used as the other ('reference') electrode since it is appropriate for predicting the behaviour of metals in dilute acids. All the figures in Table 5.2 are relative to the $H_2$ electrode which may be regarded as a 'metal', also having some tendency to dissolve in water. Metals above hydrogen in the table have a greater tendency than hydrogen to dissolve, while those below hydrogen have a smaller tendency than hydrogen to dissolve. Indeed, acidic corrosion can be regarded as 'electrolytic' corrosion with hydrogen as the other 'electrode'. In the case of zinc and copper in contact in water, electrons will flow from the zinc to the copper at their point of contact (conventional current in the reverse direction) such that the negative voltage on the zinc is partially cancelled and the zinc continues to corrode. The zinc is called the *anode*. The copper, which cannot corrode, is called the *cathode*. Hence, electrolytic corrosion will occur between any two metals in contact if

Table 5.2   Standard electrode potentials of pure metals

| Metal | Electrode potential (V) |
|---|---|
| Magnesium | –2.4 |
| Aluminium | –1.76 |
| Zinc | –0.76 |
| Chromium | –0.65 |
| Iron (ferrous) | –0.44 |
| Nickel | –0.23 |
| Tin | –0.14 |
| Lead | –0.12 |
| Hydrogen (reference) | 0.00 |
| Copper (cupric) | +0.34 |
| Silver | +0.80 |
| Gold | +1.4 |

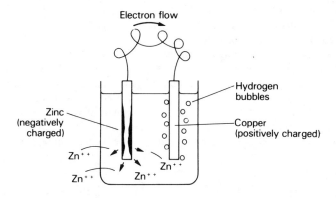

Fig. 5.9 Electrolytic corrosion resulting from zinc and copper rods immersed in aqueous solution while in electrical contact.

Table 5.3 Some common situation in which electrolytic corrosion occurs

| Situation | Metal which corrodes | Remedy |
|-----------|---------------------|--------|
| Galvanised water cistern or cylinder with copper pipes; traces of copper deposited due to water flow | Galvanised film corrodes electrolytically at point of contact with the copper particles. Film is destroyed, then steel corrodes similarly | Use a sacrificial anode in cistern. Otherwise use a plastic cistern/ copper cylinder |
| Brass plumbing fittings in certain types of water | Zinc-dezincification | Use low zinc content brass or gunmetal fittings |
| Copper ballcock soldered to brass arm | Corrosion of solder occurs in the damp atmosphere resulting in fracture of joint | Use plastic ball on ballcock |
| Copper flashing secured by steel nails | Steel corrodes rapidly | Use copper tacks for securing copper sheeting |
| Iron or steel railings set in stone plinth using lead | Steel corrodes near the base | Ensure that steel is effectively protected by paint |
| Steel radiators with copper pipes | Steel radiators corrode | Corrosion can be reduced by means of inhibitors |

moisture is present, the metal which is higher in Table 5.2 corroding. The corrosion will be more serious if the anode area is small and the cathode area large (e.g. galvanised nails in copper sheet) or if salts are present to assist ion movement in solution or if the temperature is raised. Table 5.3 gives some common situations in which electrolytic corrosion can occur.

Very important also is the fact that corrosion can occur within a piece of a single metal type, since its structure, and also possibly environmental variations, may result in effectively different electrode potentials at different parts of the surface. Table 5.4 indicates a number of ways in which this may occur. Steel readily corrodes in damp environments by mechanisms such as these. Oxygen assists in the corrosion of steel by combining with electrons from cathodes to form the hydroxyl ions necessary to produce rust (ferric hydroxide)

$$2H_2O \ + \ O_2 \ + \ \underset{\substack{\text{from}\\\text{cathode}}}{4e^-} \ \rightleftharpoons \ \underset{\substack{\text{hydroxyl}\\\text{ions}}}{4(OH)^-}$$

The corrosion of steel is also greatly increased if salts such as sodium chloride are present; a very small concentration increases the electrical conductivity of the electrolyte by many times, facilitating the flow of ions in solution. Salts may also be directly involved in the corrosion process. Electrolytic corrosion is by far the most serious form of corrosion of metals in general building applications.

Table 5.4    Situation in which electrolytic corrosion of a single metal may occur

| Cause | Anode | Examples | Remedy |
|---|---|---|---|
| Grain structure of metals | Grain boundary | Any steel component subject to dampness | Keep steel dry |
| Variations in concentration of electrolyte | Low concentration areas | All types of soil | Cathodic protection |
| Differential aeration of a metal surface | Oxygen-remote area | Improperly protected underground steel pipes | Cathodic protection |
| Dirt or scale | Dirty area (oxygen remote) | Exposure of some types of stainless steel to atmospheric dirt | Use more resistant quality or keep the surface clean |
| Stressed areas | Most heavily stressed region | Steel rivets | Protect from dampness |

# Prevention of corrosion

It will be apparent that, since both acidic and electrolytic corrosion involve ionisation, each can be prevented by keeping the metal dry, moisture being necessary for ionisation. Where dampness cannot be avoided, impermeable coatings will greatly reduce the likelihood of corrosion. However, no coating is absolutely waterproof. The resistance of a paint film, for example, increases with its thickness at these positions. Hence paint films must be the subject of regular maintenance checks, more aggressive environments calling for thicker coatings or more frequent checks. Many metals, such as aluminium, zinc, lead and copper, are protected by thin oxide or corrosion films, hence they may not need painting in clean environments. Corrosion may, however, be promoted by rain in polluted atmospheres or by contact with acid soils.

There are other means by which corrosion can be prevented or reduced – for example, galvanising of steel, in which a thin zinc coating is applied. The steel is protected for a time, even if exposed by scratches in the zinc. The zinc corrodes in the region of scratches when damp is present, supplying electrons to the steel and hence maintaining it at a sufficient negative voltage to prevent corrosion. On continued corrosion of the zinc, the exposed area of steel increases until it is too large for the zinc to give full protection (Fig. 5.10) Thick coatings, as obtained by hot dip galvanising (approximately 100 µm), are necessary if effective long-term protection is to be achieved without painting. Processes such as electroplating – often applied to articles such as nuts and bolts – give much lower thicknesses and consequently lower protection. As with other corrosion processes, the life of galvanised articles depends very much on the presence of corrosive agencies, such as pollution – in clean atmospheres, galvanised steel may last 50 years or more, but in polluted or salty environments, coatings may only last a few years.

A further possibility is to charge artificially the exposed metal negatively, using a direct current supply (Fig. 5.11). The effect is similar to galvanising except that the negative charge is supplied direct rather than by corrosion of a zinc coating. This method is used for the protection of underground steel pipelines, though it is designed to prevent corrosion occurring in defects in a

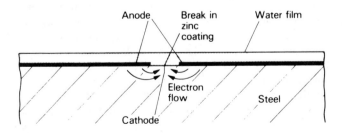

Fig. 5.10   Protection provided by galvanising at breaks in the zinc coating.

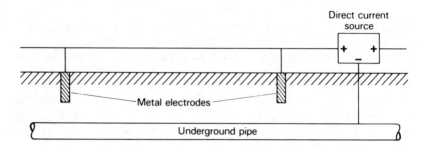

Fig. 5.11   Protection of an underground steel pipe by the impressed current method.

protective coating rather than to obviate the need for a protective film. Checking the condition of such films in a buried pipe would clearly be very difficult. The technique has also been used for protecting steel reinforcement in carbonated concrete. In such cases, it is very important to employ closely spaced electrodes in order that all embedded metal be sufficiently negatively charged relative to its *immediate* environment. In the case of reinforcing steel the 'ideal' is an electrically conductive paint applied to the concrete surface and connected to the positive side of the DC electrical supply. The reinforcing bars would be connected to the negative terminal.

## Stainless steels

These are produced by alloying steel with chromium which, if present in sufficient quantities (at least 10 per cent), forms a coherent, self-healing oxide film on the surface of the metal. Best resistance to corrosion is obtained with a single-phase material – that is, with carbon fully dissolved. This means preserving the high-temperature form of iron (see page 132) which can be achieved using nickel – a metal which, like high-temperature iron, has a face-centred cubic crystal structure. Another name for high-temperature iron is austenite; hence stainless steels containing nickel are referred to as 'austenitic' stainless steels. A typical example contains 18 per cent chromium and 11 per cent nickel – referred to as type 305 in BS 1449, Part 2, which specifies stainless steels. Welding of this steel may cause problems owing to carbon migration at high temperatures, but steels can be stabilised by the addition of niobium or titanium. Addition of molybdenum can impart even higher corrosion resistance as might be required in aggressive environments or where cleaning is not possible; for example, type 316 (BS 1449), which contains 17 per cent chromium, 11 per cent nickel and 3 per cent molybdenum.

By using a low-carbon content steel in conjunction with about 13 per cent chromium, a lower cost material with good formability can be achieved – known as ferritic stainless steel. It is used for items such as balustrades and steel sinks and can be distinguished from austenitic stainless steels because it is magnetic, whereas the latter are non-magnetic (not attacted by a magnet).

Use of a higher carbon content with 13 per cent chromium gives a harder (martensitic) stainless steel. Such steels can be used for cutlery or tools.

## Metals used in construction

Table 5.5 gives details of the composition, characteristic properties and applications in construction of a number of metals and alloys.

## Experiment

### Experiment 5.1    The effects of heat treatment on steels of differing carbon content

*Apparatus*
Samples of steel wire of low- and high-carbon content (diameters should be the same, values of 2 and 3 mm are suitable); Bunsen burners; large beaker; two pairs of pliers; tensometer (if available); small hacksaw and supply of new blades; wire cutters.

*Procedure*
1.  *Untreated properties.* Cut suitable lengths of each type of wire and compare:
    (a) Ductilities, by attempting to bend and rebend wires until failure; the number of rebends to cause failure may be noted.
    (b) Hardness, by cutting through using a hacksaw with new blade; the number of strokes to cut through may be noted.
    (c) Tensile properties. Ideally a tensometer should be used. A low carbon steel will show a definite yield point if a load extension graph is plotted. High-carbon steel wire will show no such yield point and will sustain a much higher load for a given diameter before failure (Fig. 5.4). If tensile testing apparatus is not available, the difficulty of producing a permanent bend in the two types of wire, using pliers, can be compared.
2.  *Properties of the quenched material.* Position a large beaker of cold water near two or more Bunsen burners set up so that they can be used to heat a 150 mm length of wire. Hold the low-carbon steel wire in pliers

**Table 5.5** Properties, recognition and applications of some common metals

| Metal | Type | Composition | Special Properties |
|---|---|---|---|
| Low-carbon steel | Mild steel | Carbon 0.1–0.25% Mn 0.5% | Low cost, good machinability, ductility, weldability, toughness |
| | High yield steel | Carbon 0.1–0.25% Mn 1.5% | Greater yield strength than mild steel. Ductility not generally as good as mild steel |
| | Corrosion-resistant steel (Corten) | Up to 0.5% copper. Up to 0.8% chromium | Increased resistance to corrosion – thinner sections possible |
| | Stainless steel | Contain up to 18% chromium, 10% nickel and other elements such as molybdenum | Highly resistant to corrosion. Varying hardness and ductility according to type |
| Cast iron | Grey | Up to 4% carbon, often high silica content | Low melting point – easily castable. Hard, brittle |
| Aluminium | Pure | 99.50–99.99% aluminium | Good corrosion resistance in clean atmospheres. (Protect from salt, cement mortar, contact with copper.) Can be hand-formed |
| | Alloys | Contain manganese silicon, magnesium, copper | Stronger than pure aluminium. Can be heat treated |
| Copper | – | Commercially pure | Ductile but work hardens. Noble metal, good corrosion resistance |
| Brass | Alpha or alpha-beta | Copper alloyed with up to 46% zinc | Lower zinc content brasses are soft and ductile. Higher zinc content brasses are more brittle. Corrosion resistant |
| Phosphor-bronze | – | Copper; up to 13% tin; up to 1% phosphorus | High strength and corrosion resistance. Can be hardened |
| Lead | – | Commerically pure | Eminently ductile, does not work harden. Very durable. (Protect from organic acids and cement mortar) Creeps. High thermal movement |
| Zinc | – | Commercially pure | Fairly ductile, moderate durability in clean atmospheres. Avoid contact with copper or copper alloys |

| Recognition | Application | | | |
| --- | --- | --- | --- | --- |
| | Structural | Cladding/roofing | Aesthetic | Manufactured |
| Easily identified by high density and high stiffness. Soon rusts in damp atmosphere. Mild steel reinforcing bars normally plain round | Steel sections. Reinforcing bars Wire | Plastic-coated steel sheet guttering | Wrought products | Window frames, office furniture, sundries. |
| As mild steel. High yield steel reinforcing bars are ribbed. More springy than mild steel | Steel Sections. Reinforcing bars | – | – | Nuts, bolts, springs |
| Orange/brown coating, darkening with age to purple/brown | Exposed sections | | | |
| Usually 'satin' finish. Some types may rust if dirty | Reinforcing bars | Sheet cladding | Balustrades | Tubing, sanitary applications |
| High density, rough surface texture. Brittle fracture revealing grey granular surface | Compression members in older structures | Rainwater goods | Ornamental work – staircases | Boilers, radiators, manhole covers, door furniture |
| Low density. Bright greyish metal surface when new, white deposit when corroded Fairly soft metal | – | Flashings, weatherings, guttering | – | Aluminium foil for insulating boards |
| As above | Lightweight frames | Roofing, cladding, rainwater goods according to type | – | Door handles |
| Characteristic 'copper' colour darkening to brown on ageing. Green 'patina' formed in dirty atmospheres | – | Sheet roofing, flashings, weatherings | Household utensils | Tubing |
| Yellow/gold colour, can be polished | – | – | Door furniture | Pipe fittings, hinges, screws, etc. |
| Dark copper colour | – | Widely used for fixing of claddings | – | Weather proofing strips |
| High density, dull grey finish, very easily distorted | – | Sheet roofing or cladding, flashings, weatherings | Castings such as rainwater hoppers | Pipes |
| Higher density than aluminium, fairly soft, light grey finish, becoming powdery with age | – | Roofings, flashings. weatherings | – | Door handles (die cast alloy) |

and heat the end until it glows red. 'Soak' for about one minute, then plunge the sample immediately into hot water. Repeat for the high-carbon steel wire. Test both samples as described in (1) and compare properties.

3. *Properties of the annealed material.* Repeat (2) but, instead of plunging the red hot samples into cold water, withdraw them as slowly as possible from the bunsen flame by lifting them vertically upwards. Repeat the tests of (1) on these samples.

4. *The effect of quenching followed by annealing.* To demonstrate that quenching does not chemically alter the steels (though cracking may result in objects of large section) quench samples as in (2) and then anneal as in (3) before testing. Compare results with those for the samples previously subjected to operation (3) only.

## Conclusion

Comment on the effects of each process on the properties of the low- and high-carbon steel. Note that most wires of this diameter are manufactured by drawing through a die. This produces *work hardening* so that annealing may result in reduced tensile performance compared with the untreated wire.

## Questions

1. (a) Give characteristic properties of:
   (i) iron
   (ii) cementite
   (b) Describe the physical form of pearlite.
   (c) State how the proportions of pearlite and cementite in steel depend on its carbon content.

2. Describe the essential properties of low-carbon (mild) steels.
   Explain why a high-carbon steel would not be suitable for reinforcement for concrete.

3. Describe the effects of the following elements on the properties of steel, giving typical percentages present in mild steel:

   (a) sulphur
   (b) silicon
   (c) manganese
   (d) niobium

4. (a) State, with reasons, which type of steel is more susceptible to heat treatment: low-carbon or high-carbon steel.
   (b) Describe briefly the following heat treatment processes:
   (i) annealing
   (ii) normalising
   (iii) quenching

5. (a) Explain how the cooling rate of a welded steel joint affects the performance of the weld.

   (b) Indicate the importance of the 'carbon equivalent' of a steel in relation to welding.

6. A Grade 50C steel gives a ladle analysis as follows:

   | | |
   |---|---|
   | Carbon | 0.21 per cent |
   | Manganese | 1.5 per cent |
   | Chromium | 0.025 per cent |
   | Molybdenum | 0.015 per cent |
   | Nickel | 0.04 per cent |
   | Copper | 0.04 per cent |

   Calculate the carbon equivalent of the steel and hence comment on its weldability.

7. State the three mechanisms by which metals may corrode. Indicate which of these is responsible for most damage to metals used in buildings.

8. Give three examples of situations in buildings where electrolytic corrosion may result from the use of different metals in contact.

9. Give three mechanisms by which electrolytic corrosion can occur within a single piece of metal such as steel in a damp situation.

10. Explain the function of traditional oil paints in the protection of metals from corrosion. State precautions which should be taken to ensure that paint fulfils this function effectively in practice.

11. State characteristic properties or features of the following metals which would enable them to be identified:

    (a) high-tensile steel
    (b) copper
    (c) brass
    (d) phosphor bronze
    (e) zinc

    Give applications of each in the building industry.

## References

### British Standards

BS 1449, *Steel plate, sheet and strip.*
BS 4360, *Specification for weldable structural steels.*
BS 4449: 1988, *Specification for carbon steel bars for the reinforcement of concrete.*
BS 4483: 1985, *Specification for steel fabric for the reinforcement of concrete.*

BS 4486: 1980, *Specification for hot rolled and hot rolled and processed high tensile alloy steel bars for prestressing of concrete.*

BS 5896: 1980, *Specification for high tensile steel wire and strand for prestressing of concrete.*

# Chapter 6

# Timber

Although one of the earliest materials to be used in building, timber is by no means outdated and continues to play a major part in general building, particularly in domestic dwellings and furniture. More recently, use in larger public, commercial and industrial buildings in the form of trussed rafters and glued laminated sections has increased.

Timber and timber products exhibit a number of attractive properties:

- They are natural products, hence they are a renewable resource. With properly managed forestry policies trees can be effective in reducing the carbon dioxide levels in the atmosphere, therefore reducing global warming.
- Manufacture of timber products requires much less energy than alternative materials.
- The material does not pose any disposal problem after use.
- Timber has a high strength to weight ratio.
- Timber is of aesthetically pleasing appearance.
- Timber has a low thermal conductivity, hence it has a warm 'feel'.
- The material is easily worked with hand tools.
- Strong joints can be formed with adhesives.
- Structural timber sections have good fire resistance.

Problems associated with the use of timber are:

- The material is very variable due to the effects of 'imperfections' such as knots.

- Timber is subject to movement and distortion when its moisture content changes.
- Timber is subject to fungal and insect attack.
- Untreated timber burns readily in fire and therefore constitutes a fire hazard.
- Timber is subject to creep under load.

Nevertheless, with careful selection and appropriate use, timber and timber products constitute a very valuable range of building materials.

## The structure of wood

Wood is a naturally occurring fibrous composite, the fibres consisting chiefly of crystalline cellulose $(C_6H_{10}O_5)_n$ chains of high molecular weight bonded by non-crystalline hemicellulose and lignin adhesives to form rigid, though largely hollow, cells.

These cells differ considerably from one another according to the classification of timber (whether softwood or hardwood), the particular species and their various functions within each tree.

## Softwoods

Figure 6.1 shows radial, tangential and transverse sections of a 5-year-old softwood log and it is evident that cells run chiefly in two directions – longitudinally and radially. There is a preponderance of cells running in the longitudinal direction in order to support the tree and to provide food conduction. Further food conduction and storage are provided by rays which run in the radial direction. Rays also increase the strength of the timber in this direction.

In softwoods the function of both food conduction and support are mainly fulfilled by a single type of cells known as tracheids. Additionally in the rays there are hollow food storage cells known as parenchyma. Figure 6.2(a) shows a simplified radial section of the softwood cell structure. Adjacent cells are joined in all directions by pits which are responsible for transmission of food between them. The rays in softwood are usually small and not visible to the naked eye. Figure 6.3 shows a scanning electron microscope photograph of cut transverse, radial and tangential surfaces of a sample of European redwood (*Pinus sylvestris*). Note the resin ducts which are found in some softwoods and the variation in thickness of the tracheid cell walls between spring (rapid growth) and summer (slow growth).

Softwood timbers are coniferous (cone bearing) and have needle-shaped leaves which are retained in winter. They grow relatively quickly, the resultant wood generally being soft, of low density and easily worked. Softwoods are substantially more economical than hardwoods. They are now the most widely used type of wood for general structural purposes.

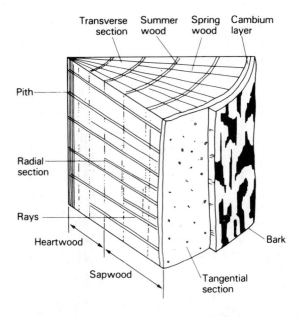

Fig. 6.1   Radial, tangential and transverse sections, through a 5-year-old softwood log.

Commonly used species in the UK are Scots pine (redwood), European spruce (whitewood), Douglas fir and cedar.

## Hardwoods

Hardwoods have a more complex cell structure than softwoods, structural support being provided by long, thick-walled cells known as fibres, while food conduction is by means of thin-walled tubular cells called vessels (Fig. 6.2(b)). Parenchyma (food storage cells) are again present both as longitudinal strands and in rays, the latter often being much more pronounced visually than in softwoods. Cells are again interconnected by pits. Sap requirements during the growing season are considerable, since new leaves must be formed. Hence vessels in springwood are often quite large and easily recognised in cross-sections of the tree. Hardwoods can be subdivided into ring-porous varieties in which the vessels are mainly concentrated in springwood and diffuse porous woods in which there is a more even distribution of vessels. Figure 6.4 shows a scanning electron microscope photograph of oak – a typical ring-porous wood. Notice also the prominence of the rays in the wood.

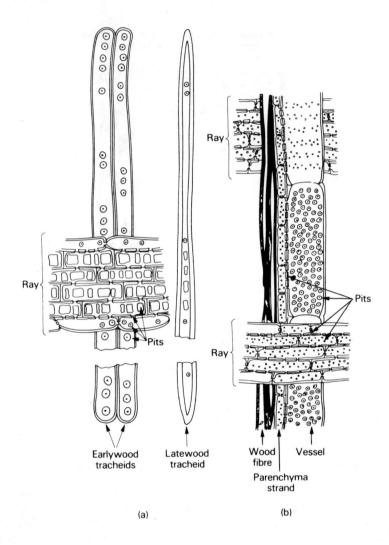

Fig. 6.2 Simplified radial sections: (a) softwood; (b) hardwood.

Hardwoods generally grow more slowly than softwoods and are usually harder, denser, stronger and more durable. There are, however, exceptions such as balsa and poplar. Hardwoods may be used where their durability is beneficial – for example in window and door sills, or in high-quality joinery. They are also extensively used in producing veneers for doors and furniture.

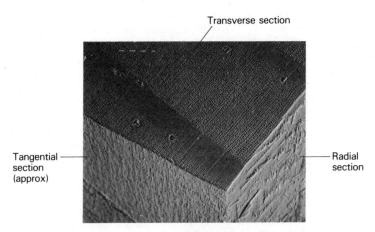

Transverse section

Tangential section (approx)

Radial section

Fig. 6.3 Scanning electron microscope photograph of European redwood (*Pinus sylvestris*), a common softwood. The resin ducts form mainly in summerwood. (×10 magnification.)

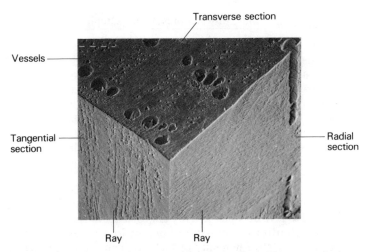

Transverse section

Vessels

Tangential section

Radial section

Ray     Ray

Fig. 6.4 Scanning electron microscope photograph of oak (*Quercus robus*) – a ring porous hardwood. There are broad rays between each group of about 4 vessels in springwood. Most of the radial section in this picture is occupied by a ray with a clear horizontal grain. (×10 magnification.)

## Heartwood and sapwood

The food conduction and storage functions of trees are fulfilled by the outer, more newly grown layers of cells, referred to as sapwood, the remainder being referred to as heartwood. The latter no longer stores food but is still important with regard to support of the tree, the cells in them tending to become acidic in order to assist preservation of the tree. Heartwood is normally darker in colour than sapwood and, on account of its acidity, tends to be more resistant to fungi and insects in service. In young trees most of the wood is sapwood, while in old trees as much as 80 per cent of the wood is heartwood. Hence, bearing in mind the difference in durability of the two types, it will be evident that older trees tend to give timber of a better quality. Traditionally sapwood was often rejected, but since this would be uneconomic when, as is now common, young trees are employed, other means of ensuring durability are now recommended, such as the use of preservatives. (See Experiment 6.1 for a detailed treatment of the identification of wood species.)

## The effects of moisture in timber

The moisture content of timber is defined as:

$$\text{Moisture content} = \frac{\text{Mass of water present in a sample}}{\text{Mass of that sample when dry}} \times 100$$

It will be noticed that, if the mass of water contained exceeds the mass of dry timber, a moisture content of over 100 per cent is obtained and this is, in fact, not unusual in newly felled (green) timber. The sapwood tends to contain more moisture at the time of felling, the average moisture content being about 130 per cent, though values of over 200 per cent are possible. In heartwood the moisture content is more likely to be around 50 per cent. The moisture content of green timber varies considerably with species; denser timbers, which include most hardwoods, have smaller cavities and larger cell walls, hence they can store less water.

The moisture content of timber in service is greatly influenced by the fact that wood is hygroscopic, that is, it tends to attract water from a damp atmosphere and to give up water to a dry atmosphere. In consequence timber will adopt an equilibrium moisture content in a given environment, depending on the relative humidity of that environment. Since increase of temperature tends to reduce relative humidity, it also results in reduced equilibrium moisture contents. Figure 6.5 shows typical values for softwoods in buildings. At moisture contents in excess of 20 per cent, timber becomes susceptible to fungal attack, hence the importance of providing a dry environment will be appreciated.

The moisture content of wood can be easily estimated by means of portable meters based on the relationship between electrical resistance and moisture content. A two-pin electrode is pushed into the wood. At lower moisture contents, electrical resistance is greatly increased. When using these meters, note the following:

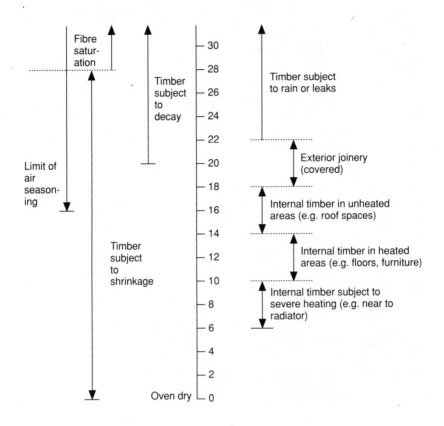

Fig. 6.5   Moisture contents of timber in various environments (per cent).

- Calibration depends on species type; readings may vary typically by 2 per cent for a given 'true' moisture content.
- The reading depends on temperature; for example; reducing the temperature by 10 °C would reduce the apparent reading by about 1.3 per cent moisture content.
- The reading obtained will be an average for the wood around and between the pins.

A further most important property of timber is that the cell walls become swollen on wetting, so that wet timber invariably occupies more space than dry timber. At moisture contents below fibre saturation value (approximately 28 per cent) the volume of timber varies linearly with its moisture content. At moisture contents over saturation value there is no variation, since additional water occupies cell cavities where there is no swelling effect. The amount of moisture movement depends on the direction considered, typical average shrinkages between the saturated state and 12 per cent moisture content being as follows:

tangential:     4–10 per cent
radial:         2/3 of tangential
longitudinal:   less than 0.1 per cent

These values reflect the fact that wood is most stable in the longitudinal direction, since most cells are aligned in this direction, while some stability in the radial direction is afforded by the rays. As would be expected, the moisture movement of timber varies considerably according to type. An important consequence of the variation of movement of timber with direction is that distortion occurs when green timber is dried. Since tangential shrinkage is greater, growth rings tend to be thrown into tension and therefore become straighter on drying. Figure 6.6 shows exaggerated sketches of the effect of drying a number of sections cut from a log.

## Seasoning

Seasoning may be defined as the controlled reduction of the moisture content of timber to a level appropriate to end use. Correctly seasoned timber should not be subject to further significant movement once in service, unless, of course, leaks occur. Seasoned timber is also immune to fungal attack, is stronger, has lower density and is therefore easier to handle or transport, or to work, glue, paint or preserve than wet timber. The importance of

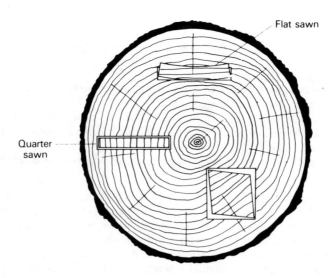

Fig. 6.6   The effect of drying various sections cut from a log.

controlled reduction of moisture will be clear when it is appreciated that moisture loss always takes place at or near the surface of the timber, hence shrinkage occurs preferentially at these positions. Surface layers are therefore thrown into tension as they contact, resulting in 'fissures' in the form of checking (longitudinal surface rupture) or splitting (longitudinal rupture passing through the piece) if drying is carried out too quickly. Serious splits in large timbers are sometimes called 'shakes'. Evaporation tends to occur more quickly from the ends of timber during seasoning, so that checking and splitting are often more pronounced at ends.

Traditionally a large amount of timber was 'air seasoned' by stacking in open formations in a roofed enclosure. Drying takes place, due to the effects of wind and sun, in a time which depends on the type and section size of timbers, the size of air gaps between them and climatic conditions. The process minimises damage but is slow, often taking several years to complete and will not reduce the moisture content below about 16 per cent, which is too high for most internal purposes. Its use in the UK has now therefore been largely been replaced by kiln seasoning in the case of softwoods, although it is still used for initial drying of hardwoods such as oak which require a low rate of drying if damage is to be avoided.

Kiln seasoning involves heating the timber in sealed chambers, initially using steam to maintain saturation, the humidity then being progressively reduced to produce drying. Since water flows more freely through the timber at high temperatures, drying can take place more rapidly without damage. Drying schedules depend on the type and size of timber sections and their initial and target moisture content. Kiln seasoning takes about one week per 25 mm thickness for softwoods and about two weeks per 25 mm thickness for hardwoods. The process has the advantage that it sterilises the wood and can reduce moisture contents to lower levels than air seasoning – values of about 12 per cent can be reached quite easily.

## Conversion of timber

This is the process by which the size of the log is reduced to a value appropriate for use and at which seasoning can be more easily carried out. A series of 'planks' is produced from each log, usually in the thickness range 25–75 mm. The cheapest and quickest way of doing this is by 'through and through cutting', as shown in Fig. 6.7(a). However, this results in some timber cut parallel to the grain ('flat sawn'), some timber cut at right angles to the grain ('quarter sawn') and some intermediate. All the non-quarter sawn timber will be more subject to distortion on drying, particularly if relatively wide, thin sheets are required, though it may be argued that flat sawn timber, having a wider grain spacing, looks more attractive. An

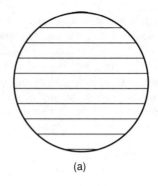 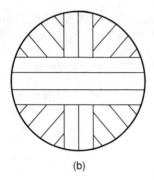

(a)                                                                    (b)

Fig. 6.7   Methods of conversion of timber: (a) 'through and through cutting'; (b) 'quarter sawn'.

alternative is to cut the timber as in Fig. 6.67b), though this is much more expensive, being slower and producing more waste than through and through cutting.

## Factors affecting the distortion of timber prior to use

These include the following (see Fig. 6.8 for explanation of terms).

- *Species of timber.* Some timbers having similar moisture movements in the radial and tangential directions – for example, Douglas fir and utile – are inherently less prone to distortion than timbers having high relative movements – for example, European redwood and beech.
- *Method of conversion.* Quarter sawn timber distorts less on moisture change than flat sawn timber, though the latter has a more interesting grain pattern. Pieces of timber which contain the pith are also more prone to distortion, especially springing. Where the piece is not cut parallel to the longitudinal axis, bowing, springing and twisting are likely.
- *Slope of grain.* Twisting is often caused when the grain is not straight or when the density of the wood varies – for example, as a result of unequal growth on different sides of the tree. Timber having uniform and straight grain is less likely to undergo distortion.
- *Stacking procedure.* Each piece of timber in a stack must be adequately supported by 'stickers' prior to and during seasoning in order to avoid distortion due to self-weight. Twisting of timber can be largely avoided if it is held firmly in stacks during seasoning.
- *Moisture changes prior to installation.* Rapid moisture changes after seasoning – for example wetting or severe drying on occupation of new properties – can lead to serious distortion. Such changes should be avoided.

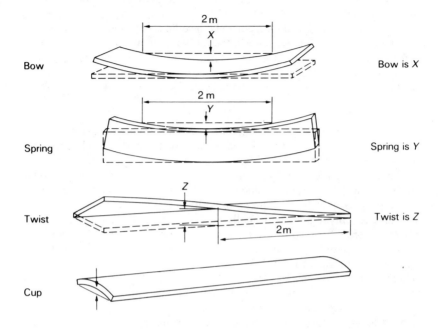

Fig. 6.8   Types of distortion in timber and their measurement.

## Durability of timber

Timber has remarkably good resistance to deterioration by atmospheric exposure. It is virtually unaffected by agencies such as rain, frost and acids, which are responsible for damage to other types of material. The main mechanisms of attack involve its use as food by plants, in the form of fungi, and by insects. In each case the destruction is brought about by enzymes which 'digest' the cellulose fibres and/or the lignin adhesive, converting them into food.

### Fungal attack

The growth of all fungi in timber requires a moisture content of at least 20 per cent, since the enzyme action necessary cannot take place in completely dry timber. There are numerous varieties, some attacking the growing tree or the unseasoned green timber, while others are able to attack the wood in service. The former are mainly the concern of the timber producer, hence attention here is given only to fungi which are active in seasoned timber. Where the cellulose only is destroyed, the wood breaks into small cubes and the rot is known as 'brown rot'. Where both cellulose and lignin are destroyed, the wood becomes soft and fibrous, the rot being known as 'white rot'. Rots can also be divided into dry rot and wet rot. There are many species falling into the latter category but they are separated from dry rot because remedial treatment of dry rot must be much more carefully carried out.

## Dry rot

(*Serpula lacrymans*, formerly called *Merulius lacrymans*.) Dry rot, a brown rot, is so named because it finally leaves the wood in a dry, friable condition. The greatest risk is in areas without satisfactory ventilation such as cavities in floors or walls, especially where there are leaks or rising damp. Growth begins when rust-red spores come into contact with damp timber, the fungus developing as branching white strands (hyphae), which form cotton-wool like patches (mycelium) and finally soft, fleshy, spore-producing, fruiting bodies (sporophore). Once established, the hyphae can grow into adjacent timber, brickwork and plaster in search of moisture or food. They may become modified into vein-like structures (rhizomorphs) 2 or 3 mm in diameter, which are able to transport moisture from damper areas to continue the attack in drier timber. Dry rot thrives at normal indoor temperatures but is killed at temperatures of over 40 °C – hence it does not grow in positions subject to solar radiation. It becomes dormant at temperatures approaching freezing. Drying out of timber renders the fungus dormant and it may die after one year in this state, though the process may take longer if the ambient temperature is reduced.

The fungus is more common in wetter parts of Great Britain, such as the north and west, though in any region it is most likely to occur in old derelict or damaged buildings which are more prone to dampness problems. London has a high incidence of dry rot outbreaks, probably for this reason.

Dry rot can be a serious problem once established, since it is often very difficult to eradicate the hyphae completely, especially in brickwork or plaster or where access is limited. Normal remedial measures are:

- Cut out and burn all visibly affected timber, together with an extra 600 mm of apparently sound neighbouring timber which may contain hyphae.
- Remove any affected plaster, again working beyond the visibly affected area.
- Preserve sound exposed timber and treat surrounding brick or concrete with a fungicide.
- Replace affected timber with preserved, seasoned timber.
- Use zinc oxychloride in paint or plaster coatings applied to affected wall surfaces. This has a fungicidal action.
- Rectify the cause of the dampness.

More recently there has been concern over the use of drastic measures to combat dry rot infestations. Experiments have been carried out into control by more environmentally friendly methods, particularly in the case of structural timbers in some older buildings as a part of conservation work. As indicated above, dry rot does not thrive in dry, well-ventilated environments so that it is possible, with careful monitoring, to destroy the fungus by controlling the environment. This also obviates the need to remove affected timber unless its strength is seriously affected. Perhaps the most important aspect of this type of protection is to carry out regular inspections; hence this technique is only feasible where timber is easily accessible for inspection or where the cost of a remote monitoring sytem is considered justifiable.

*Wet rot*

As the name suggests, wet rots require higher moisture contents in order to thrive, optimum values being in the region of 50 per cent. There are many types of wet rot, most common in the UK being *Coniophora puteana* (formerly *Coniophora cerebella*) and *Phellinus contiguus* (alternative name *Porio contigua*).

*Coniophora puteana* ('cellar fungus') is commonly found in very damp situations in buildings, especially basements or cellars. The fungus is of the brown variety and leads to cube formation, especially in large timber sections, though a skin of sound timber may be present. Surface undulations in timber may be the first sign of damage. Affected wood turns dark brown or black but fruiting bodies are rare.

*Phellinus contiguus* is responsible for a considerable increase in window joinery decay which has occurred over the last 20 years or so. It is a white rot, hence wood breaks into soft strands. The cause of the increase in this type of rot is probably the use of sapwood in joinery, which, together with poor design and inadequate glues, leads to water penetration at joints. Since most joinery of this type is painted, the first signs of attack are often undulations in or splitting of the paint film.

Damage from these fungi is more serious in the southern half of Great Britain, due to the added effect of certain wood-boring weevils which are found in association with wet rots in these regions, increasing the rate of deterioration.

The essential difference in remedial treatment of dry and wet rots is that the latter do not normally infect neighbouring dry timber, brickwork or plaster, hence it is necessary only to cut away and replace affected timber. It is nevertheless still important to remove the cause of dampness, and preservation of new timber is recommended in case any temporary dampness should recur.

## Insect attack

Insects, such as beetles that commonly infest timber in construction, have a characteristic life cycle: egg, caterpillar (larva), chrysalis and finally adult beetle. Beetles then emerge from the timber and fly to new timber to lay eggs and repeat the cycle, which may take several years. Most damage is done by larvae, which burrow through the wood leaving fine powder ('frass') behind. Flight holes are the normal visual means of locating infestations but quite advanced deterioration accompanied by serious strength loss may occur with only a small number of holes. Dust is often ejected from these holes and an indication of the degree of current activity can be judged from the colour of exit holes and dust produced – a lighter colour indicates recent attack.

Table 6.1 shows some common types of insect. The powder post beetle is mainly a problem in timber yards, while the others can thrive in indoor, seasoned timber. Note that sapwood is much more susceptible to insect attack than heartwood. The seriousness of attack also varies according to species; for example the common furniture beetle is extremely widespread but normally takes many years to produce a noticeable number of flight holes or advanced deterioration.

The house longhorn beetle is relatively rare but in parts of southern England, and particularly in west Surrey, attacks have been severe. Additionally, since the life cycle is up to 11 years in length and beetles leave a sound surface skin, serious damage can occur before many flight holes become visible. Roof timbers are mainly affected and Building Regulations require effective protection of roofing timbers to be used in certain local authority areas in these regions.

Table 6.1   Some types of insect attack on seasoned timber

| Beetle | Life cycle | Visual signs | Timber attacked |
|---|---|---|---|
| Common furniture (*Anobium punctatum*) | 3–5 years; emergence May–Aug. | Beetles 5mm long, granular dust; 0.5mm diam. flight holes | Sapwood of untreated hardwoods or softwoods |
| House longhorn (*Hylotrupes bajulus* ) | 3–11 years | A few large holes up to 10mm; surface swelling, beetles 10–20 mm in length | Sapwood of softwoods (Surrey) |
| Death watch (*Xestobium rufovillosum* ) | 4–10 years; emergence March–June | Large bun-shaped pellets; ticking during mating period, May–June | Mainly hardwoods, especially if old or decayed |
| Powder post beetle (*Lyctus species*) | 1–2 years; emergence June–Aug. | 10mm exit holes; fine dust | Sapwood of wide-pored hardwoods partly or recently seasoned |

The death watch beetle is chiefly a problem in partially decayed oak members of older churches and other historic buildings. Treatment in such cases can be difficult, due to access problems and the very large sections often affected.

## Preservative treatments for timber

Preservatives should be toxic to fungi and insects, chemically stable, able to penetrate timber and non-aggressive to surrounding materials particularly metals (BS 1282). For any given situation, the decision as to what (if any) preservative to use should be based on:

- The natural durability of the wood species being employed.
- The risk, in that situation, of damp or insect attack.
- The ease of inspection of the component.
- The consequences of attack – for example, whether failure of a component could lead to a safety hazard such as structural failure.
- The ease with which maintenance and repair work can be carried out.

Preservatives can be classified according to type:

### Tar oil preservatives (BS 144)
Creosotes are the best-known examples – they are derivatives of wood or coal, though coal distillates are superior. They are only suitable for external use, since they give rise to a noticeable colour and smell and render timber unsuitable for painting. They are, however, cheap and can be applied to timber with a fairly high moisture content. They tend to creep into porous adjacent materials such as cement renderings and plaster.

### Water-borne preservatives (BS 4072)
These consist of salts based on metals such as sodium, magnesium, zinc and copper and also arsenic and boron. Copper–chrome–arsenate (CCA) is a common example. They are tolerant of some moisture in the timber, non-flammable, non-creeping and do not stain timber, which may be painted on drying. In some varieties, such as CCA, insoluble products are deposited in the wood, resulting in excellent retention properties. Copper salts, such as CCA, colour the wood green and this is one way of confirming that the wood has been treated.

Although water-based preservatives are fairly cheap, they have the disadvantage that they swell the timber and will cause corrosion of metals in contact until drying is complete – this taking about 7 days, depending on timber type and section thickness. Hence they are best used as a preliminary protection followed by kiln drying, rather than as an *in-situ* treatment. Timber should be below fibre saturation point (28 per cent moisture content) before application of water-based preservatives. If there is a risk of timber becoming damp in use, any embedded metals should be corrosion-resistant; for example, galvanised (occasional dampness only) or stainless steel.

*Organic solvent preservatives (BS 5707)*
Examples are chloronaphthalenes, metallic naphthanates and penta-chlorophenol. Penetration is excellent provided timber is fairly dry, hence they are often preferred where only brush or spray applications are feasible. They are non-creeping, non-staining, non-corrosive to metals, non-swelling and quick-drying. Painting is usually possible, once drying is complete; usually in a few days. They are widely used for remedial treatment, though many organic solvents are highly flammable and present a fire risk until dried, especially when used in confined areas such as roof spaces. When used for new work the moisture content of the timber should be that required for end use, since the preservative does not increase the moisture content. They are often used to preserve new joinery, though it is important that all machining/drilling etc. be carried out prior to application if maximum protection is to be achieved. They are more expensive than the other types.

*Boron diffusion*
In this process freshly felled timber is dipped into a concentrated borate solution which diffuses throughout the wood by means of its own moisture, hence the need to treat the wood in a 'green state. The process can be used to advantage on 'difficult' timbers, i.e. those which do not readily admit preservatives.

## Application methods

The technique of application is just as important as the choice of a suitable preservative, since any preservative, poorly applied, will fail to penetrate well into timber sections. The difficulty increases with the size of timber sections, and some timbers such as Douglas fir have naturally low permeability to preservatives. Where machining or drilling operations are to be carried out, they should be completed prior to application of preservatives in order that the maximum penetration is obtained at all exposed surfaces. The following techniques can be used.

*Brushing or spraying*
These give only limited penetration of timber, although they may be the only feasible methods for *in-situ* remedial work. However, if several coats are applied and the treatment is repeated periodically, a useful degree of protection is achieved.

*Dipping*
This involves immersion in the preservative for a period, which, if long enough, can give considerable penetration of preservative. In the 'open tank' method of dipping, the preservative is heated and then cooled while the timber is submerged; this gives even better results. It is nevertheless not suited to hardwoods or low-permeability softwoods.

*Double vacuum process*
A vacuum is applied to timber in a sealed chamber and the preservative then

introduced. The vacuum is released, forcing the preservative into the wood. After a soaking period, the vacuum is reapplied to eject excessive preservative. This process is now widely used in conjunction with organic preservatives for external joinery such as window frames.

## Pressure impregnation

This is often used for injection of water-based and tar oil-based preservatives. In the empty cell process, the air surrounding the timber is compressed, the preservative introduced and then a still higher pressure applied, forcing preservative into the wood. On releasing the pressure, excess preservative is rejected. If the timber is not pressurised prior to injection of the preservative (in some cases, a vacuum is first applied), the retention of preservative is increased (full cell method). This method is more expensive in terms of the quantity used and, since saturated, will take longer to dry. Good protection is nevertheless achieved.

## Current preservation requirements for new building work

The Building Regulations (1991) give general requirements concerning the use of timber in situations which might cause dampness, together with specific requirements for timber used in areas in which there is a risk of attack by the house longhorn beetle.

Much more detailed requirements for timber to be used in specific situations in buildings are given in *The National House Building Council Manual* and in BS 5268, Part 5. Fundamental to such guides are knowledge of both the natural resistance of timber to decay as well as the ease with which specific timber types can be treated. The durability of timbers is classified as follows:

| | |
|---|---|
| VD | very durable |
| D | durable |
| MD | moderately durable |
| ND | non-durable |
| P | perishable |

In terms of resistance to treatment with preservatives, timbers are classified as follows:

| | |
|---|---|
| ER | extremely resistant |
| R | resistant |
| MR | moderately resistant |
| P | permeable |

Examples of classifications of some common timbers used in the UK are given in Table 6.2

It will be noticed from Table 6.2 that hardwoods are generally (though not always) more durable and more resistant to penetration of preservatives than softwoods. The table applies to heartwood only; all sapwoods are

Table 6.2   Classification of heartwoods of some common timbers in relation to durability and treatability.

| Durability Class | Treatability Class | Species | Hardwood (H) Softwood (S) |
|---|---|---|---|
| VD | ER | Iroko | H |
|  |  | Teak | H |
| D | ER | European oak | H |
|  | R | Western red cedar | S |
| MD | R | Douglas fir (UK) | S |
| ND | R | European whitewood | S |
|  | R | Sitka spruce | S |
|  | MR | European redwood | S |
| P | P | Beech | H |

regarded as non-durable or perishable. The non-durable (ND) group referred to in Table 6.2 are all commonly used for structural timber in the UK. In the NHBC standard, they would be required to be preserved:

- where there is a risk of attack by the house longhorn beetle
- in tiling battens
- in all flat roof timbers
- in timber used for external walls
- in timbers in contact with the ground or below dpc
- in external woodwork

It will be noticed that timber for use in roofs and floors is not normally required to be preserved. The risk in these situations is small, provided good construction techniques are employed.

## Timber for structural uses

The short-term tensile strength of perfect timber, taken in relation to its self-weight, compares very favourably with that steel, values of over $100 \, \text{N/mm}^2$ being common in a material which has a density equal to about 7 per cent that of steel. The compressive strength is much lower, due to the tendency of fibres to buckle on compression, but quite high bending strengths are nevertheless obtainable, timber being particularly useful for bending applications in situations where self-weight accounts for a substantial proportion of total weight – as in roofing and domestic flooring. The main problems in utilising timber efficiency are as follows:

- Strengths, even of perfect ('clear') specimens, are very variable, coefficients of variation (equal to standard deviation divided by mean strength) of over 20 per cent often being obtained.
- *Long-term* strengths are much (typically 40 per cent) lower than *short term* strengths, owing to the tendency of timber to creep.
- The strengths of individual pieces of bulk timber are greatly reduced by the presence of defects, such as knots and fissures, and by factors such as the rate of growth and slope of grain.
- Strength depends on species and, since identification is not always easy, working stresses may have to be based on poorer quality species available.

Considerable progress has been made in the last 15 years towards overcoming some of these difficulties by the introduction of techniques for grading each piece of timber according to its likely performance. Stress grading may, at the present time, be carried out visually or by machine, each method being specified in BS 4978.

## Visual stress grading

Each piece of timber is examined visually for a number of defects which would limit strength, the timber hence being categorised into one of three grades: Special Structural (SS), General Structural (GS) and Reject. The SS grade is estimated to have strength of between 50 and 60 per cent of that for clear (perfect) timber of the same species and the GS grade approximately 30 to 50 per cent. Timber should be graded by a qualified grader and stamped with the grade and mark of a recognised authority. The Timber Research and Development Association (TRADA) is the main authority in the UK.

(For experimental details, see Experiment 6.3.) The main factors to be considered are given below. Key terms relating to visible stress grading are given in Fig. 6.9 and requirements for SS and GS grades are given in Table 6.3.

### Knots
The term 'knot area ratio' (KAR) is used to estimate the weakening effect of knots, being defined as the proportion of any cross-sectional area which is occupied by knots. Since knots near to an edge have a more serious effect, a margin condition is defined as being present when more than *half* the top or bottom quarter of any section is occupied by knots (Fig. 6.10). KAR values in excess of one-third (margin condition) or one-half (no margin condition) lead to rejection of that piece.

### Fissures
These result from seasoning problems and are most common in larger sections, especially near the end of the piece. The size of a fissure is equal to the depth of cracks and the length of deeper fissures must be controlled in areas of higher bending moment, since they reduce shear resistance, and hence bending strength, parallel to the grain.

Table 6.3   BS 4978 requirements for visually stress-graded timber

| Property | | SS | GS |
|---|---|---|---|
| Knots | Margin condition | KAR $^1/_5$ or less | KAR $^1/_5$–$^1/_3$ |
| | No margin condition | KAR $^1/_3$ or less | KAR $^1/_2$–$^1/_3$ |
| Fissures | Not through the thickness | Fissures less than 10mm depth can be ignored | |
| | | Not greater than $^1/_4$ of length or 600 mm whichever is lesser | Not greater than $^1/_4$ of length or 900 mm which- is lesser |
| | Through the thickness | On any running metre, not more than one fissure of the maximum length | |
| | | Only permitted at ends with a length not greater than the width of piece | Not greater than 600 mm. If at ends with a length not greater than 1.5 × width of piece |
| Slope of Grain | | Must not exceed 1 in 10 | Must not exceed 1 in 6 |
| Wane | | Must not exceed $^1/_4$ of dimension on which it occurs for full length; $^1/_3$ in any 300 mm length | Must not exceed $^1/_3$ of dimension on which it occurs for full length; $^1/_2$ in any 300 mm length |
| Rate of growth | | Average width of annual rings not greater than 6 mm | Average width of annual rings not greater than 10 mm |
| Distortion | Bow | Thickness 35 mm not to exceed 30 mm Thickness 44 mm not to exceed 25 mm Thickness 75 mm not to exceed 10 mm Interpolate intermediate sizes | |
| | Spring | Width 60 mm not to exceed 10 mm Width 250 mm not to exceed 5 mm Interpolate intermediate sizes | |
| | Twist | Not to exceed 1mm per 25mm of width | As for SS |
| | Cup | Not to exceed $^1/_{25}$ of width, maximum 6 mm | As for SS |

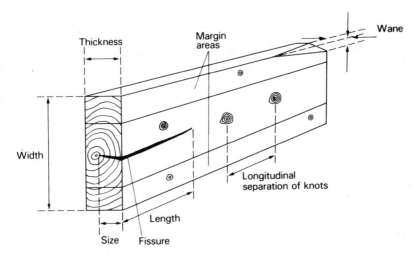

Fig. 6.9   Key terms relating to visual stress grading.

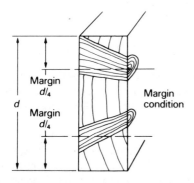

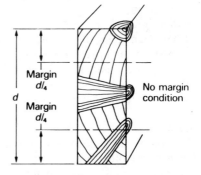

(a)  Since more than one half
of a margin area is occupied
by knots, the knot area
ratio must not exceed:

$1/3$ for GS grade
$1/5$ for SS grade

The KAR for this
section is about $1/2$ – the piece
is therefore 'reject'

(b)  Since less than one half
of a margin area is occupied
by knots, the knot area
ratio must not exceed:

$1/2$ for GS grade
$1/3$ for SS grade

The KAR for this
section is between $1/3$ and $1/2$ –
it corresponds to the GS
grade

Fig. 6.10   Knot area ratios: (a) margin condition; (b) no margin condition.

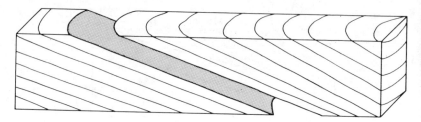

Fig. 6.11    Planes of weakness resulting when log is not cut parallel to grain direction.

## Slope of grain

This results where the tree does not grow straight or where the log is not cut parallel to the grain direction. A plane of weakness is produced where the grain intersects the surface, leading to the possibility of tensile or shear failure (Fig. 6.11).

## Wane

Wane occurs when a corner of the piece is cut too close to the outer circumference of the tree (Fig. 6.9). Its presence reduces bending strength in centre regions or bearing strength near the ends of a beam.

## Rate of growth

It has been found that faster grown softwoods have lower strength and stiffness than slower grown examples of the same species. Growth rates depend on climatic and other conditions and often vary greatly during the life of the tree but the rate of growth requirement guards against excessively fast grown timber being used. (This rule does not apply to hardwoods.)

## Distortion

The four types of distortion have already been described (Fig. 6.8). Several of these are often found in the same piece. Some types are more probable and more serious than others for a given application; for example, cupping is found in floor boards and skirting boards, while bowing, springing and twisting often occur in structural timber. Bowing and twisting are a particular nuisance in wall plates, as are springing and twisting in hip rafters. It should be noted that if timber is graded while wet, and especially if recently removed from a banded pack, significant further distortion can occur during drying.

## Other defects

These may include wormholes, sapstain and abnormal defects such as fungal decay or 'brittleheart'. See BS 4978 for guidance on such defects.

## Machine stress grading

This is based on the correlation which has been found between bending strength (modulus of rupture) and modulus of elasticity, as determined by deflection tests on samples to bending (Fig. 6.12). A major attraction of the method is that it involves a simple mechanical test on each piece rather than relying on broad relationships between strength and measured defects. The 1 per cent strength value is obtained directly from the correlation of Fig. 6.12, the only additional factor required being the factor of safety. Figure 6.12 was derived for European redwood (*Pinus sylvestris*) and European whitewood (*Picea abies*), which together form a very large proportion of the UK market at present.

Modern stress grading depends heavily on computer-operated machines which rapidly measure deflections under load of successive portions of each timber piece as it passes through. Timber is normally loaded on face rather than on edge for practical reasons and the machine automatically compensates for any out-of-straightness. A coloured dye representing the stress grade is sprayed on successive portions of the timber as it passes through. A final strip is sprayed at the end of each piece representing the worst (and therefore the final) grade for that piece. Visual inspection is also carried out for defects such as fissures, wane, resin pockets, etc. At the present time there are four machine grades: M75 and M50 (originally defined

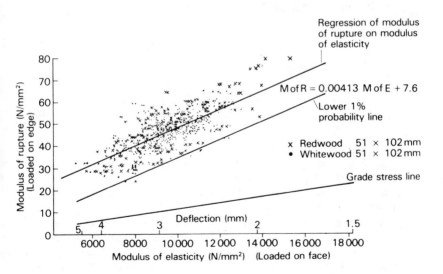

**Fig. 6.12** Correlation between elastic modulus and modulus of rupture for timber. (Crown copyright.)

in CP 112) and MSS (Machine Special Structural) and MGS (Machine General Structural) specified in BS 4978. The grades MSS and SS, MGS and GS respectively, are comparable, strengths in decreasing order being M75, MSS/SS, M50, MGS/GS. A further advantage of machine grading is that better grades are obtained than by visual grading of the same batch and there is less reject material.

Stress-grading machines must comply with an approving authority if grading in accordance with BS 4978, and machines are subject to unannounced periodic inspection in order to ensure that standards are maintained. Machine stress grading is now widely used for timber to be employed in structural elements, such as roof trusses and glued laminated timber.

### Derivation of permissible stresses (BS 5268)

The safe working stresses for any particular piece of a given species have been derived by the following process:

1.  BS 5268 stresses are based on strength tests carried out on grade SS dry timber of depth 200 mm. A special statistical distribution called the Weibull distribution was used to give failure strength levels in bending, with 5 per cent failures.
2.  Grade stresses were obtained by multiplying the characteristic stress from (1) by factors allowing for duration of load, size and factor of safety. For bending, these factors are:

    (a)  Section depth factor: $0.849 = \left[ \dfrac{200}{300} \right]^{0.4}$

    (Grade stresses are based on 300 mm sections, whereas experimental data are based on 200 mm depth sections.)

    (b)  Duration of load factor: 0.563. This gives the long-term loading stress comparative to short-term test results in which creep is much reduced.

    (c)  General safety factor: 0.724.

    These factors give an overall reduction of 0.346 (divide the characteristic stress by 2.89).
3.  For GS timber, a relativity factor, obtained by experiment, must be applied to the SS value. For bending it is 0.71.
4.  Other properties – for example, elastic moduli and compressive stresses – are obtained in the same way if experimental data are available on bulk samples. For some species, however, tests on 'clear' (perfect) samples, which previously formed the basis of design, are still used in the absence of adequate data on bulk samples, conversion factors being based on known correlations between the behaviour of small clear sections and SS grade sections.

To take an illustration for European redwood/whitewood, the SS grade characteristic bending stress, with 5 per cent failures, of a $200 \times 100 \, mm$ section is $21.6 \, N/mm^2$. Therefore,

$$\text{SS grade stress} = \frac{21.6}{2.89} = 7.5 \, N/mm^2$$

$$\text{GS grade stress will be } 7.5 \times 0.71 = 5.3 \, N/mm^2$$

These stresses are seen to be well below the short-term bending strength of clear timber (typically $70 \, N/mm^2$ ), the various causes of the large reductions having been outlined above.

## Design of structural timber sections

This is covered at present by BS 5268, which deals with a variety of structural components from solid timber as well as structures from glued laminated timber, plywood and hardboard. By way of illustration, the strength classes and the principles of design of solid timber joists and struts will be considered.

### Strength classes in BS 5268

The design procedure for structural timber can be quite complex, since the designer would need to be aware of the timber species available, together with stress grades for each species. It is not sufficient to refer to a stress grade such as 'SS' because different species have, in general, different stress capacities at a given grade. If, therefore, one species is no longer available, there may be some difficulty in finding a replacement with similar structural properties. The introduction of strength classes in BS 5268 helps alleviate such problems by grouping different species/grades together, so that the designer need no longer specify an individual species (unless there are other reasons for such a requirement). Table 6.4 shows main properties of the strength classes defined, classes 1 to 5 generally being softwoods and classes 6 to 9 being denser hardwoods. Table 6.5 shows the classification of some common species. The strength classification, as well as its stress grade, should be marked on timber in order to simplify its selection for specific purposes. By grouping strengths into classes, some species will inevitably not be used at their full grade stress and the option remains to use selected species at their appropriate stress, though for most purposes, the simplification provided by strength classes should be attractive to designers. The timber design tables in Building Regulations are based on strength classes SC3 and SC4 and form an extremely simple method of design for straightforward structures such as domestic floors and roofs

Table 6.4   Dry stress levels appropriate to BS 5268 strength classes

| BS 5268 strength class | Bending parallel to grain ($N/mm^2$) | Tension parallel to grain ($N/mm^2$) | Compression parallel to grain ($N/mm^2$) | Compression perpendicular to grain[a] ($N/mm^2$) | Shear parallel to grain ($N/mm^2$) | Modulus of elasticity[b] ($kN/mm^2$) | |
|---|---|---|---|---|---|---|---|
| | | | | | | Mean | Min. |
| SC1 | 2.8 | 2.2 | 3.5 | 2.1 | 0.46 | 6.8 | 4.5 |
| SC2 | 4.1 | 2.5 | 5.3 | 2.1 | 0.66 | 8.0 | 5.0 |
| SC3 | 5.3 | 3.2 | 6.8 | 2.2 | 0.67 | 8.8 | 5.8 |
| SC4 | 7.5 | 4.5 | 7.9 | 2.4 | 0.71 | 9.9 | 6.6 |
| SC5 | 10.0 | 6.0 | 8.7 | 2.8 | 1.00 | 10.7 | 7.1 |
| SC6 | 12.5 | 7.5 | 12.5 | 3.8 | 1.50 | 14.1 | 11.8 |
| SC7 | 15.0 | 9.0 | 14.5 | 4.4 | 1.75 | 16.2 | 13.6 |
| SC8 | 17.5 | 10.5 | 16.5 | 5.2 | 2.00 | 18.7 | 15.6 |
| SC9 | 20.5 | 12.3 | 19.5 | 6.1 | 2.25 | 21.6 | 18.0 |

[a] Compression, perpendicular to grain stresses only apply to bearing areas in the absence of wane.
[b] Mean elastic modulus figures can be used where load is shared between members – for example, flooring joists, where the boards spread the load.

## Design of joists in simple bending

The species of timber to be used must first of all be identified, the timber stress graded either visually or by machine and classified as green or dry, depending on whether the moisture content is over or under 18 per cent. The most economical result will be obtained by using pieces of the same species and grade. Supposing, for example, dry GS grade imported redwood is available. Table 6.5 shows that this timber is in strength class SC3 and Table 6.6 shows that this timber has the following properties:

Bending stress 5.3 N/mm$^2$
Mean modulus of elasticity ($E_{mean}$): 8.8 kN/mm$^2$
Minimum modulus of elasticity ($E_{min}$): 5.8 kN/mm$^2$

Suppose this timber is required to be used in a domestic floor of span 3 m in the form of joists at the normal spacing of 400 mm. It is necessary to estimate the load to be carried by the joists. The superimposed load would normally be the largest – in Building Regulations the value of 1.50 kN/m$^2$ applies in most cases. In addition, the self-weight of the floor must be calculated. For the purpose of calculation the following might be assumed:

| | |
|---|---|
| Size of joists: | 50 × 200 mm (the depth may not be correct but the error resulting will be small because the self-weight of joists is generally a small proportion of total weight) |
| Thickness of floor boards: | 20 mm |
| Density of timber: | 600 kg/m$^3$ |

Table 6.5   Some common softwoods and their BS 5268 strength classes according to stress grade

| Standard name | Strength class | | | | |
|---|---|---|---|---|---|
| | SC1 | SC2 | SC3 | SC4 | SC5 |
| *Imported* | | | | | |
| Parana pine | | | GS | SS | |
| Redwood }<br>Whitewood } | | | GS/M50 | SS | M75 |
| Western red cedar | GS | SS | | | |
| Douglas fir-larch }<br>Canada/USA } | | | GS | SS | |
| Hem-fir (Canada) | | | GS/M50 | SS | M75 |
| *British grown* | | | | | |
| Douglas fir | | GS | M50/SS | | M75 |
| Scots pine | | | GS/M50 | SS | M75 |
| European spruce }<br>Sitka spruce } | GS | M50/SS | M75 | | |

Weight of joists:

| = | 0.05 × | 0.20 × | 3 × | 600 × | 9.81 |
|---|--------|--------|-----|-------|------|
|   | (thickness) | (depth) | (span) | (density) | (grav. accn.) |
| = | 177 N |

Weight of floorboards per joist

| = | 0.02 × | 0.4 × | 3 × | 600 × | 9.81 |
|---|--------|-------|-----|-------|------|
|   | (thickness) | (depth) | (span) | (density) | (grav. accn.) |
| = | 141 N |

(The mass of a ceiling, if used, would need to be included.)

Total self-weight = 318 N or 0.318 kN

Total superimposed weight per joist = 1.50 × 0.4 × 3
(spacing) (span)

= 1.800 kN

Hence, total load supported = 1.800 + 0.318 = 2.118 kN
For normal spans the bending stress limits the load that can be carried by joists.

The bending moment ($M$) is given by:

$$M = \frac{WL}{8} \qquad W = \text{load (UD)}$$
$$L = \text{span}$$

Working in newtons and millimetres:

$$M = \frac{2118 \times 3000}{8}$$
$$= 7.94 \times 10^5 \text{ N mm}$$

Also: $M = fZ$ $\qquad Z = \text{section modulus}$ $\qquad b = \text{breadth}$
$$= \frac{bd^2}{6} \qquad d = \text{depth}$$

The bending stress, $f$, is $5.3 \text{ N/mm}^2$, though BS 5268 indicates that when four or more members act together such that the load is shared (as in boarded floor), the bending stress can be increased by 10 per cent (factor $K_8$). Hence, in this case the permissible bending stress is:

$$5.3 \times 1.1 = 5.83 \text{ N/mm}^2.$$

Therefore $Z = \dfrac{M}{f}$

$$= \frac{7.94 \times 10^5}{5.83}$$
$$= 1.36 \times 10^5 \text{ mm}^3$$

For $b = 50 \text{ mm}$

$$d = \sqrt{\left[\frac{6Z}{b}\right]}$$
$$= \sqrt{\left[\frac{6 \times 1.36 \times 10^5}{50}\right]}$$

$$= \sqrt{16\,300}$$
$$= 128 \text{ mm}$$

The standard size to cover this requirement would be 150 mm.

For larger spans, the deflection may be the limiting criterion and so this should be checked also:

$$\text{Deflection (UD load)} = \frac{5}{384} \frac{WL^3}{EI}$$

$$= \frac{5 \times 12}{384} \frac{WL^3}{Ebd^3}$$

where: $E$ = modulus of elasticity

$I$ = second moment of section = $\dfrac{bd^3}{12}$ (rectangular section)

Hence deflection $= \dfrac{60}{384} \times \dfrac{2118}{8800} \times \dfrac{3000^3}{50 \times 150^3}$

$$= 6.02 \text{ mm}$$

(Note that the $E_{\text{mean}}$ value is used here because load sharing is assumed. In other cases, or where the dead load forms a substantial proportion of the total load, $E_{\text{min}}$ should be used.)

BS 5268 limits deflection of joints to 0.003 × span – in this case 0.003 × 3000 = 9 mm. Hence in this case the joint is satisfactory.

Note: 1. Where very short spans are used under high loads, the complementary shear stress may need to be calculated and compared with the permissible value given in BS 5268.

2. The bearing area at the ends of joists should be sufficient to ensure that the bearing stress in the timber (compression perpendicular to grain) does not exceed the value allowed. See Table 6.4

## Design of struts

This is somewhat more complex than the design of joists because the most likely form of failure of a strut is by buckling, a phenomenon which depends very much on the context of use. In particular, the size of strut needed will depend on the following factors as well as on the axial load applied:

- The length of the strut since longer struts are more likely to buckle.
- The degree of restraint at the ends. The following situations are possible.
  *No restraint* – for example, if the strut rests on a flat surface.
  *Positional restraint* – for example, in a small socket.
  *Directional restraint* – for example, built into brickwork with substantial embedment of 300 mm or more.
- The application of eccentric loads or bending moments to the strut; for example, if a floor joist is supported on one side of the strut.

In order to illustrate the principle of design an example will be taken of a strut in the form of a stud partition (Fig. 6.13). It is estimated that an axial load of 10 kN is likely to act on each strut in this partition and the common section size of 100 × 50 mm will be tried using grade SC4 timber. The studs should be considered in two respects:

- As struts of length 1.2 m and thickness 50 mm, restrained in position (not direction) at each end in the horizontal direction in the plane of the paper. In case it is considered that the centres of the struts are restrained in direction, a possible mode of failure is illustrated in Fig. 6.13.
- As struts of length 2.4 m and thickness 100 mm, restrained in position (not direction) at each end in the horizontal direction at right angles to the plane of the paper.

First, the effective lengths of the strut must be found. Table 6.6 gives values for effective lengths ( BS 5268).

The effective lengths are, in the case of these struts, the same as the actual length: 1.2 m and 2.4 m in the two directions respectively.

The tendency to buckling of a strut depends on its 'slenderness ratio': the ratio of the effective length as defined above to the radius of gyration about the axis in question. For a rectangular section there is a simple relationship between radius of gyration and thickness, so that the ratio

$$\frac{\text{Effective length}}{\text{Thickness}}$$

can be used as an alternative. In this case the effective lengths are:

$$\frac{1200}{50} = 24 \text{ and in the plane of the paper, and}$$

$$\frac{2400}{100} = 24 \text{ at right angles to the plane of the paper}$$

Hence, in this case there is equal stability in each direction. Also affecting performance is the strain at which the strut is being expected to work; higher strains lead to an increased risk of instability. BS 5268 uses the inverse of this ratio which gives numbers which are easier to work with. Table 6.7 gives modification factors $(K_{12})$ by which the grade stresses given in Table 6.4 for compression parallel to the grain must be multiplied when designing struts. In this example, for SC4 timber, the grade stress is 7.9 N/mm². The elastic modulus for this strength class is 6.6 kN/mm² – the *minimum* value must be used unless there is load sharing between the struts. The ratio of elastic modulus to grade stress for this material is therefore:

$$\frac{6600}{7.9} = 835 \text{ (no units)}$$

The two figures 24 and 835 can now be referred to the $x$ and $y$ axes, respectively, of Table 6.7.

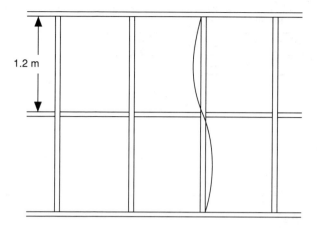

Fig. 6.13  Design of struts in a stud partition.

Table 6.6  Effective lengths for compression members

| End conditions | Effective length / Active length |
|---|---|
| Restrained at both ends in position and direction | 0.7 |
| Restrained at both ends in position and one end in direction | 0.85 |
| Restrained at both ends in position but not direction | 1.0 |
| Restrained at one end in position and direction and at the other end in direction but not position | 1.5 |
| Restrained at one end in position and in direction and free at the other end | 2.0 |

To the nearest 0.01 the table gives 0.48; hence the applied stress must not be greater than

$$0.48 \quad \times \quad 7.9 \, \text{N/mm}^2 \quad = \quad 3.8 \, \text{N/mm}^2$$

The applied stress is

$$\frac{10\,000}{100 \times 50} = 2.0 \, \text{N/mm}^2$$

This is well within the allowed stress so that the struts in the partition should be adequate.

Table 6.7   Modification factor $K_{12}$ for compression members (Based on BS 5268; not for design purposes)

| $E$ | Ratio of effective length to breadth (rectangular sections) | | | | | | | | | | |
|---|---|---|---|---|---|---|---|---|---|---|---|
| Gr. stress | 1.4 | 3.0 | 6.0 | 12.0 | 18.0 | 24.0 | 30.0 | 36.0 | 42.0 | 48.0 | 54.0 |
| 600 | 0.97 | 0.95 | 0.90 | 0.77 | 0.58 | 0.41 | 0.29 | 0.21 | 0.16 | 0.13 | 0.10 |
| 800 | 0.97 | 0.95 | 0.90 | 0.78 | 0.63 | 0.48 | 0.36 | 0.26 | 0.21 | 0.16 | 0.13 |
| 1000 | 0.98 | 0.95 | 0.90 | 0.79 | 0.66 | 0.52 | 0.41 | 0.30 | 0.24 | 0.19 | 0.16 |
| 1200 | 0.98 | 0.95 | 0.90 | 0.80 | 0.68 | 0.56 | 0.44 | 0.34 | 0.27 | 0.22 | 0.18 |
| 1400 | 0.98 | 0.95 | 0.90 | 0.80 | 0.69 | 0.58 | 0.47 | 0.37 | 0.30 | 0.24 | 0.20 |
| 1600 | 0.98 | 0.95 | 0.90 | 0.81 | 0.70 | 0.60 | 0.49 | 0.40 | 0.32 | 0.27 | 0.22 |
| 1800 | 0.98 | 0.95 | 0.90 | 0.81 | 0.71 | 0.61 | 0.51 | 0.42 | 0.34 | 0.29 | 0.24 |
| 2000 | 0.98 | 0.95 | 0.90 | 0.81 | 0.71 | 0.62 | 0.52 | 0.44 | 0.36 | 0.31 | 0.26 |

## Timber products

Timber products may offer the following advantages over solid timber:

- They are competitive in price on account of lower wastage rates; smaller sections of the tree, together with defective pieces and off-cuts which are not suitable for solid timber are often perfectly acceptable for timber products.
- Products can be produced in much larger sizes than could be obtained in solid timber – for example, sheets or long lengths.
- Defects associated with solid timber can be removed or made less significant.
- Weaknesses and distortion associated with grain can be at least partly overcome.

Areas in which timber products could cause problems are:

- The product can only be as good as the adhesive used in its production. Some products deteriorate quickly in damp conditions, as might prevail on a building site during construction, or in areas such as kitchens or bathrooms.
- Some adhesives increase wear and tear on cutting tools used for processing products.
- Some timber products are not as aesthetically attractive as their solid timber counterparts.

Table 6.8 shows some of the timber products available together with main characteristics and areas for caution. They are widely used in construction, with materials such as particle board tending to displace others such as plywood since they can be made with lower grade raw material: BS 5669

Table 6.8  Properties and applications of some timber products

| Type | Subdivision | Composition | Applications | Possible problems |
|------|-------------|-------------|--------------|-------------------|
| Fibre building boards BS 1142 | Standard hardboard | Highly compressed wood fibres; one smooth surface | Low cost internal linings; door skins | Swells in damp situations |
| | Oil tempered hardboard | Oil impregnated | External claddings | Risk of damage by impact |
| | Medium board | Lower density than hardboards | Internal linings | Protect from moisture |
| | Medium density fibreboard (MDF) | Higher strength than medium board | Decorative mouldings | Protect from moisture |
| Particle boards BS 5669 | Standard grade chipboard Type 1 | Compressed wood chips; dense surfaces | Internal claddings | Irreversible expansion in damp cond. |
| | Flooring grade Type 2 | Higher strength | Internal floors | As above |
| | Moisture-resistant flooring grade Type2/3 | Water resistant | Kitchens/bathrooms | |
| Plywood BS 6566 | Shutter ply | Crossed coarse veneers; water-proof adhesive | Formwork for concrete. Carcassing in timber frame construction | Limited re-use; |
| | Birch ply | Quality veneers waterproof adhesive | External linings, e.g. soffit boards | Good veneers are expensive |
| Glulam BS 4169 | Straight | 45 mm laminates, high-performance adhesives | Large section straight structural members | Protect from rain and sun |
| | Curved | Approx. 20 mm laminates | Curved members; portal frames | As above |

now includes a specification for the structural use of chipboard. Some other materials have recently been introduced – for example, 'wafer board', which comprises thin slices of wood bonded together; and LVL (laminated veneer lumbar), which comprises strips of wood bonded together to form structural size sections. When the benefits of cost and performance are included, such materials have a distinct advantage over solid equivalents.

## Experiments

### Experiment 6.1   Identification of common woods

Identification of a given piece of wood, especially a softwood, can be extremely difficult since there are many hundreds of species obtainable and properties within each species often vary. The following may nevertheless assist as a guide to identifying some of the commonly available softwoods and hardwoods.

It is usually essential to have a smooth surface sample of the wood to be identified – this could be in the form of a small rectangular piece. One radial and one tangential face should be planed smooth and one transverse section cut smooth with a *very sharp* knife or chisel. With careful cutting, a clean, dark surface, free of tearing, should be obtainable. Sanding will also produce a smooth surface but this takes longer and the dust tends to obscure finer detail such as resin ducts. Moistening the end of the timber may help in finding these. Low power magnification of, for example, 10 times will assist greatly in seeing and identifying detail in transverse sections. It may also be an advantage for some assessments such as knot distribution to have a larger piece of timber as well as the smaller sample.

A distinction should first be made between hardwoods and softwoods. On the transverse plane, softwoods can be recognised by the absence of pores – see, for example, Fig. 6.14 (Douglas fir).

Hardwoods can be further subdivided according to the nature of pores – whether ring-porous or diffuse-porous (Figs 6.4 and 6.17). The density of dry specimens should also be measured by weighing and dividing by volume. Tables 6.9, 6.10 and 6.11 give characteristic descriptions of the key properties of common softwoods, ring-porous hardwoods and diffuse porous hardwoods, respectively. Figures 6.14 to 6.17 show transverse sections of a number of common woods and these may assist in identifying the species concerned. Reference to scanning electron microscope photographs of Figs 6.3 and 6.4 may also be helpful.

### Experiment 6.2   Measurement of the moisture movement of timber

This can be quite easily carried out, provided suitable size specimens are available. Ideally three pieces of timber of each species should be available of size approximately $6 \times 75 \times 100$ mm, each cut with its long dimension parallel to one of the three respective principal directions in the wood (Fig. 6.18). Knot-free samples are preferable and the end grain of the flat-sawn piece should be as straight as possible if cupping is to be avoided. Pieces containing the pith should, in any case, be avoided.

Note that quarter-sawn specimens are fragile and must therefore be handled with care. An airtight cabinet with sufficient space for all the

Fig. 6.14 Transverse section of Douglas fir (*Pseudotsuga taxifolia*) – softwood. Note dense summerwood bands (×8 magnification).

specimens and glass containers of sulphuric acid are also required, together with vernier calipers accurate to 0.1 mm.

### Procedure

The sample should first be conditioned to a given low relative humidity – say, 30 per cent, which would need sulphuric acid of relative density 1.43 in a small airtight cabinet (take care!). Equilibrium may take some days to reach and samples should be weighed periodically until mass becomes constant. Mark pieces at a convenient position on the longest dimension and take caliper readings at this position. Readings should be taken several times to ensure repeatability and the mass of sample noted.

The samples are then returned and the humidity increased to, say, 90 per cent, which would need sulphuric acid of relative density 1.14. Again weigh samples periodically until the weight is constant. Soaking to 'speed up' the operation is not recommended as it leads to moisture gradients and may damage the sample.

Table 6.9    Softwood Identification (Species given in approximate order of darkening colour)

| Type | Growth rings | Density $(kg/m^3)$ | Colour | Other features |
|------|-------------|--------------------|--------|----------------|
| European whitewood (European spruce) *Picea abies* (Fig. 6.15) | Well-defined springwood merges gradually into summer-wood | 510 | Very light yellow/white | A few resin ducts irregularly spaced in late wood |
| Sitka spruce *Picea stichensis* | Springwood merges gradually into thin bands of dark red/brown summerwood | 400 | Very light brown | Irregular resin ducts – late wood; rapid growth – rings well spaced – low density |
| European redwood (Scots pine) *Pinus sylvestris* (Fig. 6.3) | Well defined. Thin/medium bands of dark red summerwood | 520 | Sapwood light brown. Heartwood deeper red/brown | Newly cut surface smells resinous. Regular resin ducts. Knots in groups separated by clear timber |
| Parana pine *Araucaria augustifolia* | Poorly defined. Early wood merges gradually into late wood | 530 | Sapwood light, heartwood light to dark brown with red streaks | No resin ducts. Dense, fine, straight grain |
| Western hemlock *Tsuga haterophylla* | Abrupt transition from springwood to pink/brown summerwood | 500 | Yellow/brown | Resin ducts absent |
| Douglas fir *Pseudotsuga taxifolia* (Fig. 6.14) | Well-defined springwood. Quite thick well-defined bands of dense hard red/brown summerwood difficult to cut. Growth rings often wavy | 545 | Darker than most softwoods except cedar. Sapwood yellow, heartwood red/brown | Resin pockets visible on transverse section under microscope. May be resinous smell |
| Western red cedar *Thuja Plicata* | Clear rings. Narrow bands of late wood | 370 | Sapwood yellow, heartwood fairly dark brown | Distinctive aromatic smell. No resin ducts. Note the low density |

Table 6.10   Identifying features of some common ring porous hardwoods

| Type | Growth rings/rays | Density | Colour (kg/m²) | Other features |
|------|------|------|------|------|
| Ash *Fraxinus excelsior* | Growth rings distinct. Fairly coarse pores in springwood. Hard, dense summerwood. Rays invisible | 690 | Very light-whitish | Quite hard, dense wood |
| Oak *Quercus robus* (Fig. 6.4) | Rays very clear. Large pores in springwood | 690 | Light yellow/brown | 'Silver grain' on quarter-sawn surfaces (effect of rays). Hard, dense |
| Sweet chestnut *Castanea sativa* | Well-defined thin rings of large pores. Rays indistinct | 600 | Light/medium brown | Similar to oak if flat sawn. Silver grain absent when quarter-sawn. Less dense that oak |

Table 6.11   Identifying features of some common diffuse porous hardwoods

| Type | Growth rings/rays | Density (kg/m²) | Colour | Other features |
|------|------|------|------|------|
| Beech *Fagus sylvatica* (Fig. 6.16) | Thin, well-defined rays on transverse section. Grain fairly faint. Pores not visible except under microscope | 710 | Light yellow/pink | Rays produce fine, well-distributed lines on flat sawn section |
| Maple *Acer saccharum* | Summerwood distinct though only slightly darker than springwood. Rays distinct. Fine, well-scattered pores | 690 | Light biscuit brown | Fine smooth surface; attractive lustre. Hard, dense |
| African mahogany *Khaya ivorensis* (Fig. 6.17) | Well-shaped medium-size pores. Thin rays – clearly visible under microscope | 660 | Dark red/brown | Pores produce long fine indentations on radial and tangential surfaces |

Fig. 6.15   Transverse section of whitewood (*Picea abies*) – softwood. Note the small, light-coloured resin ducts irregularly spaced in latewood (×8 magnification).

Fig. 6.16   Transverse section of beech (*Fagus sylvatica*) – diffuse porous hardwood. Note prominent rays (×8 magnification).

Fig. 6.17  Transverse section of African mahogany (*Khaya ivorensis*) – diffuse porous hardwood. Note the well-spaced pores and this rays. The growth rings are indistinct in this sample (×8 magnification).

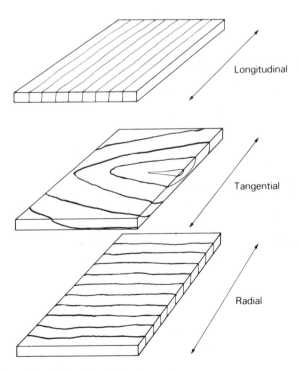

Fig. 6.18  Methods of cutting timber for moisture movement measurement.

When constant weight is achieved the caliper readings and masses are again taken. Find the difference between the readings corresponding to the two humidities and represent the movement as a percentage. (Note that the difference in readings in the longitudinal direction will be very small.)

Finally, oven dry the timber at 105 °C and find the dry mass. Hence find the moisture contents of the timber at the two humidities. Compile a table giving radial, tangential and longitudinal moisture movements between the moisture contents concerned for each species.

## Experiment 6.3   Visual stress grading of timber (BS 4978)

The basic technique of visual stress grading is very simple, though the process may be found to be time-consuming until experience is gained as to the critical criteria for any one piece. It is suggested that at least three pieces be obtained of nominal size 100 × 50 mm and length 2.0 m – this minimum length being necessary for distortion measurements. Pieces should ideally be selected so that at least one sample of each grade SS, GS and Reject is included. A simple scribe for measuring slope of grain is required - this comprises a cranked rod with a swivel handle and an inclined needle at its end (Fig. 6.19). A 3 m straightedge is also required.

*Procedure*
*Knots.* Find the worst knot or combination of knots in the piece and mark positions of surface intercepts on a sketch of the timber section, drawn full size on graph paper. Join the intercepts, inspecting the grain at the end of the piece to ascertain radial directions, if necessary. Hence obtain the knot area ratio. Note also whether a margin condition is present (Fig. 6.10).

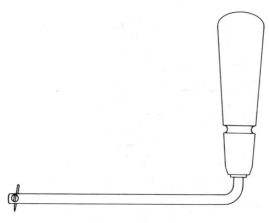

Fig. 6.19   Scribe with swivel handle for measuring slope of grain.

*Fissures.* Inspect for fissures and record:

- position
- size (depth) as a fraction of the thickness of the piece
- length

*Slope of grain.* This should be measured on one face and edge. Press the needle of the scribe into the wood and draw along in the apparent direction of the grain. The needle will form a groove and give the precise direction. Avoid excess pressure which would prevent steady movement, or inadequate pressure which would cause the needle to 'jump' across the fibres. Note that growth ring direction and grain direction do not normally coincide. Measure the slope of grain as in Fig. 6.20. The direction of fissures, when present, also indicates grain direction.

*Wane.* The wane on any face is equal to the amount of that dimension which is missing. Express this as a fraction of the dimensions concerned. For example, if wane reduces the thickness of a 100 × 50 mm piece from 50 to 40 mm, the wane is 10 mm or one-fifth of the dimension. Note the position of the wane.

*Rate of growth.* Measure the average width of growth rings at right-angles to growth, measuring rings over a 75 mm length if possible and omitting material within 25 mm of the pith.

*Distortion.* Cupping will not normally be a problem in pieces of section 100 × 50 mm. Measure twisting by weighting one end of the piece flat on the floor, such that the other end of the timber lifts. Measure the difference in uplift between the two edges of the piece (both ends will lift if the timber is bowed). Bowing and springing are easily measured using a 2 m straightedge (Fig. 6.8).

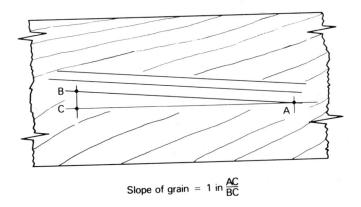

Slope of grain = 1 in $\frac{AC}{BC}$

Fig. 6.20  Method of measurement of slope of grain.

*Assessment of grade*
Compare results in respect of knots, fissures, slope of grain, wane, rate of growth and distortion with the requirements of Table 6.3. To comply with a given grade, pieces must satisfy all requirements. Any piece failing a GS requirement is designated REJECT.

## Experiment 6.4   Machine stress grading

This experiment is not an attempt to simulate commercial machine grading tests in which performance of each localised section of the piece is assessed. It should be possible, however, to correlate results of this simple test with visual grading results, especially where the latter grades are determined on the basis of rate of growth or knot characteristics. For this reason it is recommended that the experiment be carried out on pieces of timber which have already been visually graded.

It is, of course, unnecessary to test timber to destruction – deflection tests are adequate, though destructive tests will yield bending stress results which can be compared with the stresses given in Table 6.4.

Pieces should be loaded on edge; for 2 m pieces of 100 × 50 mm softwood, this will require a machine with a load capacity in the region of 10 kN. A compression machine with a suitable steel beam insert could be used if a flexural machine is not available.

*Procedure*
Locate the piece centrally in the machine (Fig. 6.21) using 50 × 50 × 6 mm steel bearers where timber touches loading and support rollers. Fix a dial gauge under the piece at centre span.

Apply the load progressively and record the load for each millimetre increase of deflection up to 10 mm. Remove the dial gauge and, if required, load the piece to destruction. Note the mode of failure.

Repeat for the other pieces.

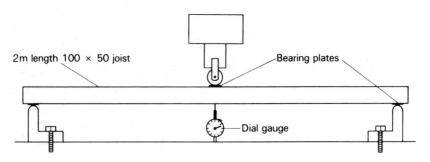

Fig. 6.21   Loading arrangement for machine stress grading.

*Results*

Plot graphs of load ($y$ axis) against deflection ($x$ axis) and draw the best straight line through the points. Obtain the gradient of the graph in N/mm. Note that higher gradients signify stiffer material.

The elastic modulus of the timber may be calculated:

$$E = \frac{WL^3}{48\delta I} \qquad (\delta = \text{deflection})$$

(other symbols as used earlier)

$$= \frac{W}{\delta}\frac{L^3}{48I}$$

$$= (\text{Gradient of graph}) \times \frac{L^3}{48I}$$

The stress at failure of the timber is

$$f = \frac{M}{Z} = \frac{WL}{8Z}$$

Calculate the stress at failure. Look for correlations between the elastic modulus and strength values for each piece and the visual stress grade result.

## Questions

1. Describe the nature and function of the following:
   (a) vessels (hardwood)
   (b) fibres (hardwood)
   (c) tracheids (softwood)
   (d) rays
2. (a) Describe the visual differences between hardwood and softwood by inspection of Figs 6.3 and 6.4.
   (b) Distinguish between ring-porous and diffuse-porous hardwoods and give one common example of each.
3. Comment on the strength, durability and availability of softwoods compared with hardwoods. Hence give main applications of each group in construction.
4. Explain the meaning of the term *seasoning*. Give three advantages of seasoned timber over wet or 'green' timber. Explain why timber for internal use should be dried to a lower moisture content than roofing timber.
5. Describe the basic causes of distortion in timber. Suggest steps by which this distortion can be minimised in practice.

6. Describe three types of fungus attack in timber, giving situations where they commonly occur and visible effects on timber. Outline methods by which the risk of decay can be minimised.

7. Describe what action should be taken when dry rot is found in the joists supporting a suspended ground floor.

8. Define the terms *knot area ratio* (KAR) and *margin condition*. Give the BS 4978 requirements for KAR values for 'General Structure' and 'Special Structural' grades. Estimate the KAR values of the sections shown in Fig. 6.22 and hence classify them in this respect.

9. Explain the principle of machine stress grading. Give two advantages of machine stress grading over visual stress grading.

10. (a) A uniformly distributed load of total value 25 kN is to be supported by 10 timber joists spanning 3 m. A joist size of $175 \times 50$ mm has been suggested. If the permissible bending stress in the timber is 5.3 N/mm$^2$, calculate whether this size of joist would be satisfactory, from a bending stress standpoint. Ignore the self-weight of the joists.

    (b) If the elastic modulus of the timber is 8600 N/mm$^2$, calculate the maximum deflection and check whether it is within the permissible limit of $0.003 \times$ span.

11. A loft conversion is proposed which will result in loading on a stud partition at first floor level. The partition is of height 3 m and comprises $100 \times 45$ mm sections at 400 mm centres with bracing at mid-height by timber of the same section. A load of 5 kN per metre run is anticipated. Calculate whether the partition would be of sufficient strength if grade SC2 timber is used (see Table 6.4).

## References

*Building Regulations* (1976) HMSO.
*National House Building Council Manual.*

### British Standards

BS 144: 1973, *Coal tar creosote for the preservation of timber.*
BS 1142: 1989, *Specification for fibre building boards.*
BS 373: 1957, *Testing small, clear specimens of timber.*
BS 1282: 1975, *Guide to the choice, use and application of wood preservatives.*
BS 4072: 1974, *Wood preservation by means of water-borne copper/ chrome/arsenic compositions.*
BS 4169: 1988, *Specification for the manufacture of glued laminated structural timber members.*

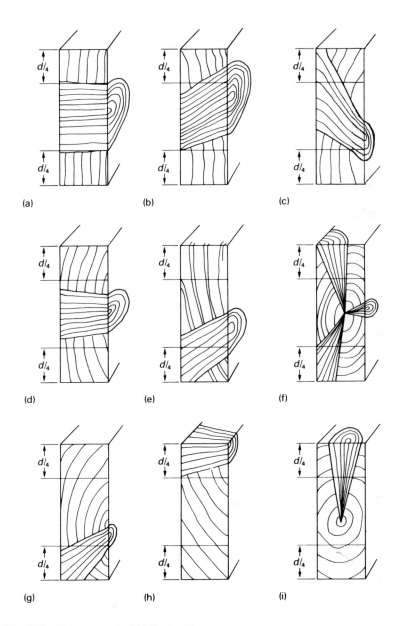

Fig. 6.22  Knot area ratio (KAR) classification.

BS 4978: 1988, *Specification for softwood timber grades for structural use.*
BS 5268, *Structural use of timber.*
BS 5669, *Particle boards.*
BS 5707, *Solutions of wood preservation in organic solvents.*

# Chapter 7

# Bituminous materials

General usage of the term 'bituminous' covers products based on tar as well as those based on bitumen, though the bitumen-based products are the more widely employed in all forms of construction. This chapter is related mainly to bituminous materials as so defined but includes others which, though strictly non-bituminous, are related in origin or use – for example, certain macadams as used in road construction.

This group of materials offers the following general attractions:

- excellent resistance to absorption/passage of water
- good adhesion to many materials
- resistance to dilute acids and alkalis
- good flexibility at normal temperatures
- properties can be varied to suit application

Problems/disadvantages associated with them are:

- tendency to become brittle at low temperatures and soft at high temperatures
- subject to creep
- have low inherent stiffness, so must be carefully blended with harder materials where stresses are encountered

**Bitumen**

Bitumen (BS 3690) consists essentially of hydrocarbons of varying molecular size which are soluble in carbon disulphide. The larger molecular weight fractions are of solid nature and these are dispersed in the lower molecular weight fractions which are either of a resinous or an oily nature.

Bitumens occur naturally in the form of asphalts, which are mixtures of bitumen, minerals and water found in the form of rock or lake asphalt. The majority of bitumen is, however, produced as a residue from the fractional distillation of crude oil.

The properties of bitumen vary greatly according to composition which, in turn, depends on the manufacturing method and crude material. However, all bitumens are thermoplastic - they soften on heating, though they have no well-defined melting point. On cooling they tend to become progressively more brittle. They dissolve in many organic solvents, though this may not always be an advantage.

## Production process

The crude oil is initially heated in a distillation column to between 300 and 350 °C as a result of which volatile fractions vaporise and are removed, producing naphtha, kerosine and gas oil according to the position in the distillation column. The remaining material, known as the 'long' residue, must be further processed to produce bitumen and this is done by heating in a partial vacuum in order to obtain further vaporisation without raising the temperature to over 400 °C, which would cause decomposition of the residue. The resulting 'short' residue may be modified by air blowing, carried out at a temperature of about 300 °C. The air has a number of effects, but overall the molecular weight of the 'asphaltenes' in the bitumen is increased and this reduces the susceptibility of the material to softening at high temperatures for a given 'penetration' (viscosity). The amount of air blowing can be varied, but road grades are usually subjected to a limited amount of this treatment.

The essential feature of bitumens is that they are solid at ordinary temperatures and must therefore be heated prior to application, whether in roofing, flooring or roads. They are nevertheless widely employed since they have the advantage that hardening takes place immediately on cooling, so that there is only a minimal delay between construction and use. For grading purposes blown bitumens are described by their penetration and/or their softening point. Penetration is the distance that a standard shaped needle will penetrate a sample of bitumen when loaded in a standard manner (100 g load for 5 s at 25 °C). The penetration is measured in units of 0.1 mm, e.g. '200 pen' bitumen means 20 mm penetration (see Experiment 7.1). Penetrations for most purposes lie in the range 30 to 300. The softening point temperature is found by a ring and ball test (see Experiment 7.4). When bitumens have a penetration of less than, say, 100, the softening point test tends to give a more sensitive indication of hardness, the softening point increasing from about 45 °C for '100 pen' bitumens to over 60 °C for very low pen bitumens. There is, however, no unique relationship between these two parameters and bitumen specifications often refer to both.

## Cutback bitumens

These usually comprise straight-run bitumens to which a volatile fluxing oil such as kerosene or creosote is added in order to reduce viscosity. The handling and placing of products based on cutback bitumens can take place at much lower temperatures than straight-run bitumens. Subsequent hardening is by solvent evaporation, hence it is generally considerably slower than that of straight-run bitumens. Their main application is in surface dressings. Cutbacks are graded using the standard tar viscometer in which the time in seconds for 50 ml of cutback to flow through a standard orifice at a standard temperature is measured (Experiment 7.3). Values range typically from 50 to 200 s.

## Bitumen emulsions

A bitumen emulsion contains minute bitumen particles – around 1 µm in size, dispersed in water by means of an emulsifying agent. These agents impart electric charges to the surface of the particles, thereby causing them to repel and prevent the formation of a continuous solid mass. Two types of emulsifier may be used: anionic emulsifiers, which impart a negative charge to the bitumen, and cationic emulsifiers which impart a positive charge.

The mechanism of solidification ('breaking') of emulsions commences when the emulsion contacts the mineral aggregate, which begins to inactivate the emulsifier by water absorption. In the case of cationic emulsions, breaking may be assisted by the neutralisation of the positive charge by negative charges often present on solid mineral materials such as silica. Hence, cationic emulsions tend to break more quickly than anionic emulsions and may break without drying of the water. Anionic emulsions are nevertheless cheaper and are more commonly used. Breaking is normally accompanied by a change in colour from brown to black.

Bitumen emulsions are widely used to produce damp-proof membranes, as curing membranes in concrete roadbases, as tack-coats in road surfacing and in surface dressings. They have the advantage of adhering to damp surfaces, though 'breaking' in such situations would normally be delayed. Thin films only must be used - ponding of an emulsion will result in very long drying times. Where thicker coatings, such as in damp-proofing, are required, these should comprise several coats, each being allowed to dry before application of the next.

## Road tar

Road tar (BS 76) is obtained by the destructive distillation of coal (or, less commonly, wood or shale) in the absence of air at temperatures up to 1000 °C. The tar is driven off as a thick brown vapour, which is condensed and collected. The other products of distillation are gas, and oil fractions such as benzole.

The condensate is referred to as crude tar and this must be treated to produce road tars. Treatment comprises distillation at temperatures up to 350 °C in which light oil, naphthalene and creosote are progressively driven off. The remaining material (residue) is described as base tar or pitch and this is fluxed back with selected tar oils to give a road tar with the desired viscosity and distillation characteristics.

According to BS 76 there are two types of road tar: type S and type C. Type S tars contain more volatile oils and therefore tend to set more rapidly on the road, being used mainly in Surface dressings. Type C tars are less volatile and are intended for use with Coated macadam. Tars are classified principally according to their viscosity as measured by the standard tar viscometer (Experiment 7.2). Water content is also important, BS 76 requiring a maximum of 0.5 per cent in order to prevent frothing on heating.

Tars differ from bitumens in the following main respects:

• They are generally more susceptible to temperature change, tending to soften quite quickly at higher temperatures and being subject to embrittlement at very low temperatures. Hence the viscosity of tar which is chosen for a particular function will depend on seasonal temperature variations as well as traffic load requirements.
• Tars and pitches undergo 'weathering' in service – a combination of oxidation and evaporation of volatile fractions. This is an advantage in roads since smooth tar films, brought to the road surface by heavy traffic, weather away slowly and leave the aggregate exposed.
• The solubility of tars in organic solvents (such as petrol) is much lower than that of bitumen, hence tars are more suitable for surfacings which are subject to contamination by diesel or lubricating oils – such as parking areas for buses or commercial vehicles.

The main application of tar is as a surface dressing for roads, since it has the advantage over bitumens (with the exception of cationic emulsions) of superior adhesion in damp conditions. Dense tar surfacings, which are rather similar to dense bitumen macadams, are also used in situations where the advantages of a tar binder are required.

Overall, the use of tar has reduced since the early 1970s when North Sea gas replaced 'town' gas – the coal used for the latter being the main source of tar. An important source of tar is now the iron and steel industry which employs coke in the furnaces. The material is also obtained as a by-product of smokeless fuels.

## Low-temperature tar

This term is misleading, since it may erroneously be taken to imply a tar for cold application. In fact the term refers to the method of production; a carbonising temperature of 600–750 °C being employed instead of 1000 °C, as for conventional tars. The lower temperature is used when coal is converted to smokeless fuels as distinct from the higher temperature 'destructive' distillation which was formerly used for the production of 'town' gas. This can be blended to give very similar properties to those of conventional tars. BS 76 makes provision for the use of such tars.

## Principles of design of road pavements

Road pavements are designed to fulfil the following functions:

(a) Provide a smooth skid-resistant surface.
(b) Reduce the stresses produced by the wheels of vehicles to levels which can be withstood by the natural underlying material (subgrade).
(c) Protect the subgrade from water penetration and from frost (which would weaken the material).

Figure 7.1 shows the basic components of a road pavement. The subgrade is the natural material occurring at that position, though it may be adjusted to give the 'formation' level upon which the road sits.

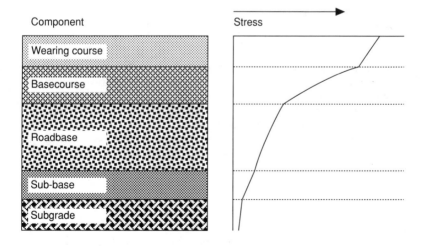

Fig. 7.1   Components of a road pavement together with principles of stress reduction.

The sub-base will normally be a granular material, assisting in functions (b) and (c) above, as well as providing a working platform for construction traffic.

The roadbase performs the main structural function. It may employ a cement binder, a bituminous binder, or (less commonly at present) simply crushed stone.

The surfacing may be of concrete, or, more commonly, a bituminous material. The latter will be in two layers: a basecourse to provide a suitable platform for the other layer, and a wearing course which provides the wearing surface and normally sheds water from the road.

The road pavement as a whole reduces wheel load stresses to acceptable levels in the subgrade, the rate of stress reduction depending on the relative stiffnesses of the various layers. Figure 7.1 shows schematically how stresses might be reduced. The subject is, however, highly complex, since excessive stress in any one layer must be avoided, the layers must work together and, most important, fatigue failure, resulting from repeated application of wheel loads, must be avoided. In addition, the layers must be able to withstand movements resulting from temperature or other effects.

## Asphalts and macadams

Asphalts are bituminous mixtures containing substantial amounts of fine material in the form of fine aggregate and filler (Fig. 7.2), which, together with the binder, form a stiff 'mortar' which provides strength and stiffness in the final product. They are described as 'gap graded' since there is little material between the sizes 2.36 and 10 mm. They employ a relatively hard bitumen in quite high proportions to produce an impermeable product with low void content and high durability

Macadams are materials which rely mainly on particle interlock rather than a stiff mortar to provide strength and stiffness. They normally contain a wide range of particle sizes in a continuous grading (Fig. 7.2), voids in the material being filled by particles of smaller size as in concrete. Macadams coated with a bituminous binder are described as 'coated' macadams. The bitumen acts as a lubricant during compaction as well as a binder. Macadams use smaller quantities of bitumen than asphalts – typically 4–6 per cent – hence they are slightly cheaper than asphalts.

Some macadams for roadbase construction do not have a bituminous binder and are referred to as 'wet-mix' and 'dry-bound' macadams.

The principal forms of asphalts and macadams are given below.

### Hot-rolled asphalts (BS 594)

These comprise a mixture of aggregate, filler and 'asphaltic cement', which would generally be a 50 'pen' bitumen. They may be used in roadbases, base courses and wearing courses.

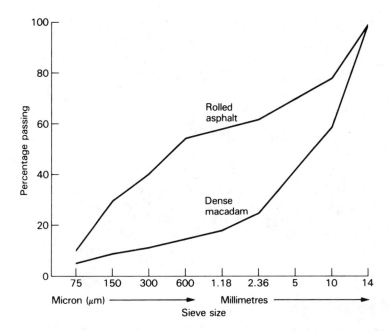

Fig. 7.2 Overall aggregate gradings for typical rolled asphalt and dense bitumen macadam. The rolled asphalt contains some coarse material and a high proportion of fine material. The macadam is composed mainly of coarse material.

Roadbases and base courses employ high stone contents, typically 50–60 per cent, and the aggregate is of relatively large size such as 14 or 28 mm. They are laid in thicknesses up to 150 mm.

Wearing courses contain higher binder contents and lower stone contents, typically 30 per cent, of smaller maximum size such as 10 or 14 mm and are laid 40–50 mm thick. Precoated chippings of size 14 or 20 mm are normally rolled in after laying to provide surface texture and hence skid resistance (Fig. 7.3).

The optimum binder content in a rolled asphalt may be obtained by the Marshall test (see Experiment 7.12) or, alternatively, 'recipe' type mixes can be used.

Hot-rolled asphalts have low permeability and high durability and are widely used on heavily trafficked roads and city streets.

## Mastic asphalts (BS 1446, 1447)

These are high-quality surfacing materials with a relatively large binder content – between 6 and 20 per cent of a bitumen of low penetration, typically between 15 and 25. The aggregate may be natural rock asphalt (BS

Fig. 7.3 Hot-rolled asphalt containing 20 mm precoated chippings, rolled into a dense impermeable mortar. The coin – a new 10p piece – gives the scale.

1446) or, more commonly, limestone (BS 1447), each ground to a powder which largely passes a 2.36 mm sieve. Relatively high application temperatures in the region of 200 °C are necessary and the material is spread by hand – a process involving a good deal of skill, the material being applied in layers to a final thickness in the range 25–50 mm. The product contains less than 1 per cent voids and is impermeable.

An important application is in flat roofing. The asphalt is laid in two layers on isolating felt to allow movement and is held in place by its own weight. The substrate for the asphalt should be rigid – for example, concrete, rather than timber, since asphalts tend to crack under excessive flexing, especially after ageing. Solar reflective treatments help reduce embrittlement due to ageing. Mastic asphalt roofs provide much better resistance to pedestrian traffic than built-up felt roofs especially if suitably gritted. They will also withstand vehicle traffic provided support is adequate. Mastic asphalt is also used for flooring and tanking.

### Bitumen macadams (BS 4987)

These are based on bitumens of 100–300 penetration, together with graded

aggregates of size up to 40 mm depending on thickness. They have lower
fines contents leading to a higher void content than asphalts (up to 25 per
cent depending on type). This reduces progressively due to compaction by
traffic. Macadams have good workability owing to their low fines content
and are easier to lay than hot-rolled asphalts.

The texture of a macadam depends on the maximum aggregate size and
on the actual content of fines – the amount of material passing a 3.35 mm
sieve. There are several types:

*Open-graded macadams*
These have a maximum size of 10–14 mm and and fines content not
exceeding 25 per cent. They can be used as basecourses (20 mm aggregate)
and wearing courses (10 or 14 mm aggregate), though since they are
permeable, an impermeable surface dressing would be needed when used as
a wearing course. Alternatively, the impervious layer can be positioned
*underneath* the wearing course, which when used in conjunction with an
open-graded macadam of aggregate size 20 mm, leads to *pervious macadam*,
greatly reducing spray in wet weather as well as giving a quieter ride (Fig.
7.4). In view of their safety advantages, trials are currently being conducted

Fig. 7.4   Pervious macadam, showing the open texture, leading to free draining
through the weaving course. The material shown here (15 mm) is widely used on
French autoroutes.

on pervious macadams to determine suitability for heavily trafficked roads though open-graded macadams are not generally considered suitable for the heaviest traffic densities.

### Medium-graded macadams

These have a maximum size of 6 mm with 45–65 per cent fines and can be used for non-trafficked areas such as footways.

### Dense macadams

Dense macadams have a fines content of about 37 per cent and typically 28 or 40 mm maximum size. The high fines results in low void contents of between 5 and 10 per cent, so that they are suitable for roadbases and basecourses for heavily trafficked roads. The 1988 edition of BS 4987 refers to dense macadam wearing courses of aggregate size 10 and 14 mm and described as *'close-graded' macadam* and introduces a *new dense wearing course* of aggregate size 6 mm, of lower void content. These materials are nevertheless not completely impervious and do not have the durability or skid resistance required for roads subject to the heaviest traffic loads.

### Fine-graded macadam

This, formerly called 'cold asphalt', is based on high penetration or heavily cutback bitumens and is designed to be used cold or warm. Fine and coarse varieties based on 6 and 10 mm aggregates, respectively, are available. Fine varieties are laid in thicknesses of about 20 mm and coarse varieties to 30 mm. They are used for regulating and patching purposes and for footpaths, having the advantage that they can be stored for some time (Fig. 7.5). They are initially soft and permeable but gradually improve in both respects on trafficking. They make little contribution to the strength of the pavement and will be indented by sustained point loads.

## Wet-mix and dry-bound macadams

Wet-mix macadam is based on crushed rock or slag graded from 40 mm downwards and mixed with 2–6 per cent of water in order to minimise segregation and to assist compaction. Compaction is carried out using a vibrating roller and the final void content should be less than 10 per cent. Wet-mix macadam may be used as a roadbase material for flexible roads.

Dry-bound macadam is also used for roadbases and comprises 40 or 50 mm nominal single size material. It is laid dry to a depth of 75–100 mm and compacted by a 2.5 tonne steel-tyred roller. Dry fine aggregate to size 5 mm downwards is then spread about 25 mm thick on the compacted stone and vibrated into the voids using a plate or roller. The process is repeated until no further fine aggregate will penetrate and the surface is then swept clean and finally rolled. Over-vibration must be avoided, otherwise segregation will occur.

Wet-mix macadam suffers from the disadvantage that accurate metering of water is necessary, while the dry-bound macadam process is more involved and requires preliminary drying of materials.

Fig. 7.5  Fine-graded macadam (laid cold), used for lightly trafficked areas such as footpaths.

The load-carrying capacity of roads constructed from each of these materials in the UK is now limited by current Road Works Specifications and there is evidence that the rate of deformation in dry-bound roadbases in particular is comparatively rapid. They are therefore no longer in common use.

## Surface dressings

These form a quick, convenient and economic means of arresting surface deterioration of a road surface. The procedure involves cleaning off loose debris by brushing, spraying a bituminous emulsion or cutback bitumen onto the dry road surface, applying chippings to the wet binder and then rolling to embed the chippings. The choice of chipping size must take into account the existing surface texture and the hardness of the existing binder. An important part of the operation is to allow traffic to use the dressed surface, but with a strictly controlled maximum speed, to avoid wheel traction tearing the new surface. The traffic helps to embed the chippings. Surface dressing is only really feasible in summer months, when the temperature is high enough to promote drying/solvent loss, and it will not arrest a deteriorating road structure.

## Defects in bituminous pavements or surfacings

### Cracking

Cracking is likely to lead to serious problems in all types of surfacing. Cracked surfacings permit water penetration, which, in the case of roads, may lead to weakening of foundations and hence reduced loadbearing capacity. The penetration of water will also accelerate deterioration due to the effect of frost and de-icing salts. In roofs, moisture penetration leads to all the associated problems of dampness both in the roof structure and in the building itself.

Cracking is caused essentially by movements which exceed those the materials can absorb. In flexible roads laid on cement-bound bases, a pattern of well-spaced cracks frequently forms in the bituminous surfacing, corresponding to cracks previously formed in the roadbase. These 'reflected' cracks can be minimised if, during construction, measures are taken which will result in a relatively large number of fine cracks in the base, so that movements at any one crack are reduced. Thicker or more flexible surfacings also reduce the extent of the problem. Once such cracking patterns are produced they are difficult to cure permanently but are relatively easy to seal, and resurfacing will effect a short-term remedy. Similar types of crack may occur in asphalt roofs where the isolation membrane between the surfacing and deck is absent or ineffective. Joints in decking, together with dressings, flashings and weatherings, require careful design and execution to avoid excessive movements leading to cracks. Where a few well-defined cracks exist, it may be possible to cut back, make provision for movement and apply fresh asphalt. Where significant water penetration has occurred, however, complete replacement will be necessary.

Finer cracks or crazing may result when a bituminous surfacing is not able to absorb thermal movements to which it is subjected. This may, in the first instance, be due to the use of a bitumen or tar having too low a penetration value, though a further cause is often overheating prior to use, which results in the loss of fluxing oils. Solar radiation causes further progressive loss of these oils and roofs which are prone to rapid heating and cooling benefit from protective coatings of reflective material such as light-coloured chippings. Crazing is a surface effect and does not, initially, lead to water penetration though it accelerates overall failure of the material, since cracks concentrate the tensile stresses occurring when cold weather causes thermal contraction. Most crazing occurs in cold weather, since this results in tensile stress at a time when the tensile strain capacity of the material is at its minimum. Surface crazing may be at least temporarily alleviated by means of surface treatments, which also require protection from solar radiation. Deeper crazing, resulting in moisture penetration, will necessitate replacement of the material.

Cracking may also result from overloading of the material either due to excessively high stresses caused by vehicle overloading or because the road foundation is inadequate. The latter may be the result of water penetration, for example, which increases stresses in the bituminous layers because of inadequate stiffness.

## Blistering

This occurs mainly in mastic asphalt or built-up felt roofs and is caused by the presence of moisture under the surface. The blistering occurs in hot weather when heat vaporises the water, which swells the softened mastic. Frequently the blisters remain on subsequent cooling – for example, if rain stiffens the asphalt and prevents it resettling. Blisters are initially waterproof but may eventually leak in cold weather if damaged mechanically. The asphalt is also invariably thinner at these positions, increasing vulnerability. Minor blisters may be left, or patched in dry weather. Severe blistering is indicative of a design or construction fault which calls for replacement. The problem occurs most commonly with 'wet' substrates such as concrete which are not effectively isolated by a membrane from the asphalt. By the use of such a membrane, together with dry conditions during construction, blistering should be avoidable.

## Deformation

Surface deformation in roads is the result of prolonged or severe mechanical stress. It is quite common in deceleration areas of roads such as roundabouts, traffic lights or bus stops, especially on downward gradients, being worst in the wheel tracks of heavily trafficked roads, especially in hot weather. The cause is the use of either too much binder, too soft a binder or incorrect aggregate grading and it leads to corrugations of the surfacing. These then exacerbate the problem as the upward-sloping sections of each undulation become subject to higher stresses under a given vehicle load. To effect a cure it is necessary to replace the surfacing with one designed to have the extra stiffness required.

A more serious form of deformation may occur in which the roadbase is also affected. This is caused by structural failure of the pavement as a whole and results in large depressions in the road surface, often in the wheel tracks of heavy vehicles in the 'slow' lane of multi-lane carriageways. In such situations the surfacing and roadbase must be removed and replaced to an uprated specification.

## Embedment

Embedment is a term commonly used in the context of road surfacings and it describes the sinking of chippings into the binder and, in some cases, the underlying material. Particularly important is surface dressing where the binder and chipping size employed should be such that, after compaction by traffic, the lower half of each surface chipping is embedded (Fig. 7.6(a)). Where embedment is less than this (Fig. 7.6(b)), there will be a tendency for chippings to break loose ('scabbing'). Conversely, excessive use of binder, combined with soft substrate, may lead to total embedment and hence to 'fatting-up' with loss of surface texture and skid resistance (Fig. 7.6(c)). Where this occurs, a further surface dressing can be applied with relatively low binder content, together with increased chipping size (up to 20 mm) in order to result in the correct degree of embedment after compaction by traffic.

Embedment can also lead to loss of surface texture and hence high-speed skid-resistance in hot-rolled asphalts.

## Experiments: tests on bitumen/tar binders

### Introductory comment

These tests are based on British Standard specifications, although some have been abbreviated or simplified to take account of the limited time usually available. Further simplifications can be undertaken where the principle only

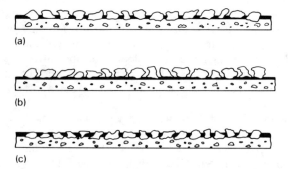

(a)

(b)

(c)

Fig. 7.6   Varying degrees of embedment of chippings applied as surface dresing: (a) satisfactory – approximately half each stone covered; (b) too low leading to 'scabbing'; (c) too high leading to 'fatting-up'.

of the test is being illustrated, though the experimenter should refer to the text of BS specifications where an accurate and valid test result is required.

## Experiment 7.1 Determination of the penetration of bitumen (BS 2000, Part 49)

This test is used to determine the consistency of penetration grade 'straight-run' and stiffer 'cutback' bitumens at the standard temperature of 25 °C.

*Apparatus*
Penetration apparatus as shown in Fig. 7.7. The dial gauge should be capable of reading to 0.1 mm and the mass of the needle and spindle assembly should total 100 g. Needles should be of hardened stainless steel, of 1 mm diameter with a 9° taper at one end to a truncated tip 0.15 mm in diameter. Sample containers should be of 55 mm diameter and 35 mm internal depth for penetrations less than 200 or 70 mm diameter and 45 mm internal depth for

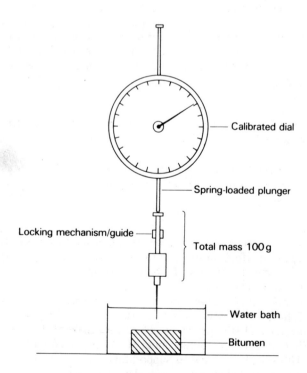

Fig. 7.7 Penetration apparatus.

penetrations between 200 and 350. They may be of metal or glass. A water bath capable of maintaining a temperature of 25 ± 0.1 °C, stop-watch and thermometer are also required.

## Preparation of specimens

Heat the sample until just sufficiently fluid to pour, stirring to avoid local overheating. Stirring also helps to remove air bubbles which may be present in blown bitumens. Avoid prolonged heating (over 30 min). Pour into the sample container judged to be appropriate to the original consistency of the material, tapping to remove air. The container should be almost full. Allow to cool for 1 to 2 h, according to container size, and then insert into the water bath containing at least 10 litres of water and leave for a similar time to allow the temperature to reach 25 °C.

## Procedure

The test should be carried out on the specimen while submerged in the water bath or while submerged in a suitable smaller vessel containing water at 25 °C. In each case check that the sample is firmly mounted and unable to rock. Lower the clean penetrometer needle until it just touches the bitumen. This may be facilitated by using the shadow of the needle or its reflected image. Take the reading of the dial gauge. Release the needle and start the stop-watch. After 5 s note the reading of the needle. The penetration is equal to the difference between the two readings expressed in units of 0.1 mm. Make three determinations, the simplest method being to use a fresh clean needle at a different part of the surface for each determination. Find the average reading.

## Experiment 7.2   Determination of the equiviscous temperature (e.v.t.) of tar using the standard tar viscometer (BS 76)

This test is designed for softer bituminous materials such as cutback bitumen. The e.v.t. is the temperature at which the viscosity of the materials is a standard value.

## Apparatus

Standard tar viscometer (Fig. 7.8). The essential features are as follows:

*Tar cup*, in the form of an open-topped brass cylinder with an accurate opening of diameter 10 mm at the base. This opening can be closed by a sphere attached to an operating rod.

*Water bath*, which contains an accurate central sleeve into which the tar cup fits. It has three legs which are adjustable for levelling, a run-off cock, electric heater and stirring vanes. The vanes are attached via a further concentric cylinder to a curved upper shield provided with a handle for rotating. This shield also provides support for the tar cup valve.

A 100 ml graduated measuring cylinder, thermometer and stop-watch are also required.

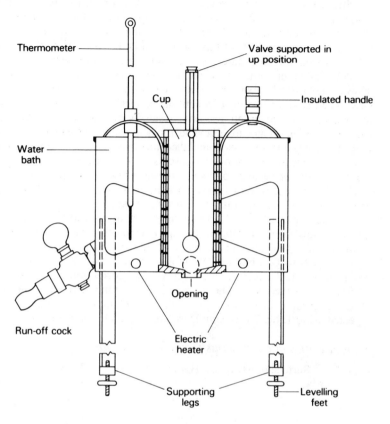

Fig. 7.8   Section through standard tar viscometer.

## Sample preparation

Immerse a container of the sample in water at a temperature of not more than 30 °C above the expected equiviscous temperature. If no information is available as to the expected e.v.t., a preparation temperature of 65 °C may be used as a starting point, heating further, as necessary, to obtain adequate fluidity. BS 76 requires the warm sample to be strained through a 600 μm sieve to remove any material which may interfere with flow through the standard orifice. Stir the tar and keep covered while hot to avoid loss of volatile oils.

## Procedure

Adjust the water bath so that the upper rim of the sleeve is horizontal, using the spirit level and the adjusting screws on the legs. Fit the stirrer assembly and fill the water bath to a level 10 mm from the top. Place the tar cap and valve in the sleeve.

The e.v.t. should be 'guessed'; if no information is available, a temperature of 35 °C may be taken. The temperature in the water bath should then be adjusted to 0.05 °C higher than the required temperature and maintained within 0.05 °C of this value (that is, between 0 and 0.1 °C higher than the required temperature) throughout the test. Stir the water frequently.

Pour 250 ml of the tar at the temperature of preparation into a beaker and allow to cool to 0.5 °C above the required test temperature. Close the orifice in the tar cup and pour tar into the cup until the correct level (indicated by a peg) is reached. Place the graduated measuring receiving cylinder under the tar cup and add 20 ml of soap solution to this cylinder to help prevent tar sticking to it. Stir the tar gently, using a thermometer, until the temperature is at the required test value and then withdraw and remove any excess tar from the thermometer. Lift the valve clear of the tar and commence timing when the meniscus of the soap reaches the 25 ml mark. Stop timing when the 75 ml mark is reached. Replace the valve. Empty the receiver to avoid tar sticking to the sides.

The time recorded will be that for 50 ml of tar (discounting the first 25 ml) to flow from the cup. At the equiviscous temperature, the time taken would be 50 s. For times in the range of 33–75 s, Table 7.1 will enable the correct e.v.t. to be obtained. Suppose, for example, a time of 65 s was obtained at a test temperature of 35 °C. The correction from the table is +1.6 °C and so the e.v.t. value is 36.6 °C. If the time obtained is outside the values indicated in heavy type in the table, the experiment should be repeated using Table 7.1 as a guide to select a new test temperature.

*Typical values of equiviscous temperature for tars*

| Surface dressing | Heavy traffic | 42–46 °C |
| | Average roads | 38–42 °C |
| | Light traffic | 34 °C |
| Coated materials | Dense roadbase (heavy traffic) | 54 °C |
| | Basecourse, estate roads | 38 °C |

## Experiment 7.3   Determination of the viscosity of cutback bitumen (standard tar viscometer BS 2000, Part 72)

The apparatus for this test is basically the same as that for Experiment 7.2. There are some differences in procedure. The cup size should generally be 10 mm and the test temperature 40 °C, though a 4 mm cup and 25 °C temperature may be used for very fluid samples.

A trial test may be necessary to determine the correct cup size and temperature, although most cutbacks have viscosities in the range 50–200 s.

### Procedure
According to BS 2000, the temperature of the cutback should be adjusted to within 0.1 °C of the test temperature by use of a separate water bath. This

Table 7.1 Correction in °C to be applied to temperatures of test to give the equiviscous temperature (e.v.t.) for tars

| Viscosity (s) | 0 | 1 | 2 | 3 | 4 | 5 | 6 | 7 | 8 | 9 |
|---|---|---|---|---|---|---|---|---|---|---|
| 10 | −10.4 | −9.8 | −9.2 | −8.7 | −8.2 | −7.7 | −7.3 | −6.9 | −6.5 | −6.1 |
| 20 | −5.7 | −5.4 | −5.1 | −4.8 | −4.5 | −4.3 | −4.0 | −3.8 | −3.5 | −3.3 |
| 30 | −3.1 | −2.9 | −2.7 | **−2.5** | **−2.3** | **−2.2** | **−2.0** | **−1.9** | **−1.7** | **−1.5** |
| 40 | **−1.4** | **−1.2** | **−1.1** | **−0.9** | **−0.8** | **−0.6** | **−0.5** | **−0.4** | **−0.3** | **−0.1** |
| 50 | **0** | **+0.1** | **+0.2** | **+0.3** | **+0.5** | **+0.6** | **+0.7** | **+0.8** | **+0.9** | **+1.0** |
| 60 | **+1.1** | **+1.2** | **+1.3** | **+1.4** | **+1.5** | **+1.6** | **+1.7** | **+1.7** | **+1.8** | **+1.9** |
| 70 | **+2.0** | **+2.1** | **+2.2** | **+2.3** | **+2.3** | **+2.4** | +2.5 | +2.5 | +2.6 | +2.7 |
| 80 | +2.8 | +2.8 | +2.9 | +3.0 | +3.0 | +3.1 | +3.1 | +3.2 | +3.3 | +3.3 |
| 90 | +3.4 | +3.5 | +3.5 | +3.6 | +3.6 | +3.7 | +3.7 | +3.8 | +3.9 | +3.9 |
| 100 | +4.0 | +4.0 | +4.1 | +4.1 | +4.2 | +4.2 | +4.3 | +4.3 | +4.4 | +4.4 |
| 110 | +4.5 | +4.6 | +4.6 | +4.7 | +4.7 | +4.8 | +4.8 | +4.9 | +4.9 | +5.0 |
| 120 | +5.0 | +5.1 | +5.1 | +5.2 | +5.2 | +5.2 | +5.3 | +5.3 | +5.4 | +5.4 |
| 130 | +5.5 | +5.5 | +5.5 | +5.6 | +5.6 | +5.7 | +5.7 | +5.7 | +5.8 | +5.8 |
| 140 | +5.9 | +5.9 | +6.0 | +6.0 | +6.0 | +6.1 | +6.1 | +6.1 | +6.2 | +6.2 |

*Note:* That part of the table giving corrections for tars having viscosities between 33 s and 75 s inclusive (indicated by bold type) may alone be used in calculating the e.v.t. to be reported. The remainder of the table will be useful in ranging tests.

requires the orifice of the viscometer to be closed by a cork inserted from beneath and the thermometer and valve to be located by a second cork with a central hole (for the thermometer) and slot fitted into the top. The cutback should be free from water and should be first heated to 60 °C for a period given below depending on sample size, taking care to avoid loss of volatile constituents.

| *Vol of sample* (ml) | *Period of heating* (h) |
|---|---|
| 750 ± 250 | 2 ± 0.25 |
| 1500 ± 500 | 2.5 ± 0.5 |
| 2500 ± 500 | 3 ± 0.5 |

Then cool to the approximate test temperature. After fitting the corks, fill the viscometer cup with the cutback until it is level with the peg. Immerse in the separate water bath for 1.5 h at the correct temperature.

Set up the viscometer water bath and adjust to the correct temperature, stirring frequently.

Transfer the viscometer cup to the viscometer water bath and remove the thermometer and corks. Check that the level of cutback is correct.

Time 50 ml of cutback through the orifice, as described in Experiment 7.2 except that light mineral oil is used in the graduated receiver instead of soap solution. Hence obtain the viscosity of the cutback expressed in seconds.

## Experiment 7.4  Determination of the softening point of bitumen using the ring-and-ball method (BS 2000, Part 58)

This test is designed to be used with 'straight-run' (penetration grade) bitumens.

### Apparatus

The apparatus enables tests to be carried out on two samples of the material. It comprises a support frame 25 mm above a lower plate, together with two tapered rings, two ball-centring guides and two steel balls.

A heat-resistant glass container, 300 μm sieve, thermometer, electric heater, mechanical (electric or magnetic) stirrer, thermometer and knife are also required. Distilled water is normally suitable as the heating medium; glycerol/dextrin debonding mixture (or a suitable grease).

The assembly is shown in Fig. 7.9.

### Sample preparation

Heat the sample until fluid; with the exception of low penetration bitumens, a temperature of 130 °C should suffice. Stir until free of air bubbles and filter, if necessary, through a 300 μm sieve (though this should normally not be necessary). Heat the tapered rings to about the same temperature and place on

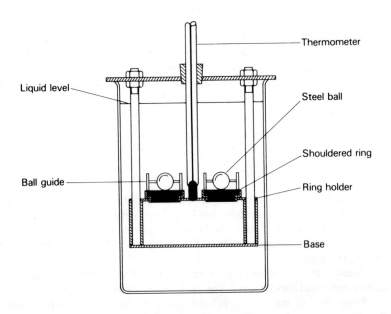

Fig. 7.9  Softening point apparatus.

a metal plate coated with a debonding mixture of glycerol and dextrin. Fill the rings with a *slight* excess of the liquid bitumen but avoid overflowing. Cool in air for 30 min and cut away the excess with a warm knife.

### Procedure (for softening points of 80 °C and below)

Assemble the apparatus as in Fig. 7.9, with ball guides in position, and fill the glass container to 50 mm above the upper surface of the rings with freshly boiled distilled water at a temperature of 5 °C. After 15 min place the steel balls, which should be at a similar temperature, in the ball guides. Heat the bath, stirring constantly so that the temperature rises at 5 ± 0.5 °C per minute. Record the temperature at which each sample surrounding the ball touches the bottom plate. The temperatures obtained from the two samples should be within 1 °C of each other. Obtain the average temperature – this is the softening temperature.

### Typical softening temperatures

| Penetration value | 35 | 50 | 100 | 300 |
|---|---|---|---|---|
| Approx. softening temperature (°C) | 58 | 53 | 46 | 34 |

(It should be emphasised that the relationship is approximate and that many specifications for bitumens require both penetration *and* softening point determinations.)

## Experiment 7.5   Determination of the water content of a tar (BS 76)

Excess water in tar can cause frothing at high temperatures and may affect adhesion.

### Apparatus

A Dean and Stark reflux apparatus with 2 ml receiver is required (Fig. 7.10). A glass or metal flask of 500 ml capacity is used as the distillation vessel; 100 ml graduated measuring cylinder; coal tar solvent naphtha; weighing balance accurate to 0.5 g.

### Procedure

Weigh 100 ± 0.5 g of the stirred tar sample into the flask and add 100 ml of solvent.

Attach the flask to the Dean and Stark apparatus and heat so that condensate falls from the end of the condenser at two to five drops per second. Continue until water is no longer visible in the apparatus, except in the bottom of the calibrated tube. The water level should also become constant.

Allow the apparatus to cool and note the volume of water. The mass of the water in grams can be taken to be the same as its volume in millilitres. Express the mass as a percentage of the original mass of tar.

*Note.* BS 76 limits the amount of water present to 0.5 per cent.

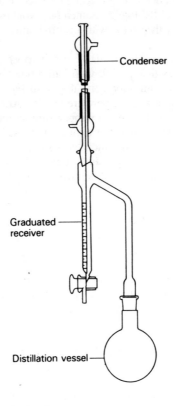

Fig. 7.10 Dean and Stark reflux apparatus.

### Experiment 7.6 Determination of the solubility of a bituminous binder (BS 2000, Part 47)

It is important to have a knowledge of the proportion of insoluble material in the binder when analysing coated materials, especially asphalts.

*Apparatus*
Filtration flask, suitable for vacuum filtering; sintered glass crucible, no. 4 porosity, effective diameter approximately 30 mm; rubber adaptor for holding the crucible in the filter flask; oven; powdered glass – Pyrex, particle diameter less than 74 µm ; balance accurate to 0.05 g; trichloroethylene solvent. This is toxic, hence adequate ventilation is necessary. All sources of heat should be kept away from trichloroethylene, since it emits phosgene gas on heating.

## Specimen preparation

If the sample contains water, it should be heated to 120–130 °C with constant stirring until the binder ceases to foam.

## Procedure

Weigh 3 ± 0.1 g of dried powdered glass into the sintered glass crucible. Dry the crucible in the oven under a slight vacuum for 1 h. Then dry to constant weight. Assemble the filter apparatus, tapping to ensure even distribution of the glass powder. Apply light suction and wash with trichloroethylene.

Weigh 2–5 g. of the dry binder to the nearest 0.01 g ($W_1$) and place in a 200 ml flask. Add 100 ml of solvent. Stir and leave for 1 h. Pour the contents into the crucible and filter gently. The filtrate should be clear. Wash the residue in the flask into the crucible with the solvent in a wash bottle. Wash the material in the crucible with further solvent until the filtrate is colourless. Dry the crucible in an oven at 105–115 °C for 1 h; cool in a desiccator and weigh the residue ($W_2$).

The solubility is then

$$\left(\frac{W_1 - W_2}{W_1}\right) \times 100$$

*Note:* The *in*soluble residue in bituminous binders for roads should not exceed 0.5 per cent.

## Experiment 7.7 Determination of the flash point of bitumen (BS 4689)

It is essential that the flash point of bitumens be sufficiently high to avoid the risk of combustion of bituminous mixtures at working temperatures.

## Apparatus

The Cleveland cup apparatus is used (Fig. 7.11); this comprises an open test cup mounted on a heating plate and held in a frame to which a thermometer and test flame applicator are also accurately fixed. The apparatus may be heated by bunsen burner. To avoid objectionable fumes, the apparatus may be mounted in a fume cupboard, but draughts must be avoided. The thermometer should be of standard type range, –6 to 400 °C, graduated in 2 °C intervals.

## Procedure

Position the thermometer bulb 6 mm from the bottom of the clean cup and towards the opposite side from the test flame burner arm. Heat the sample as necessary to make it sufficiently fluid and then fill the cup so that the meniscus is level with the top. Light the test flame and adjust to a diameter of 4 mm. Heat the sample at approx. 15 °C per minute and as it approaches the flash point (as indicated by smoke emission) decrease the heating rate to

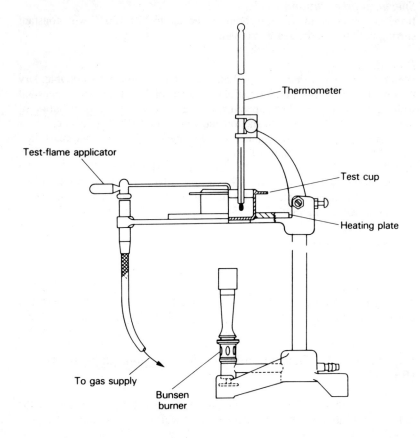

Fig. 7.11 Cleveland open cup apparatus.

about 5 °C per minute. At about 30 °C below the flash point pass the test flame across the surface of the bitumen, about 1 mm above it, taking about 1 s to pass the surface. Repeat for each 2 °C temperature rise until a flash appears on the surface of the bitumen. The temperature at this point is the flash temperature.

The heating procedure may require a 'trial run' in order to obtain correct heating rates and the point at which the test flame is applied.

## Experiments: tests on bitumen/tar mixtures (BS 598)

### Sampling and sample reduction

It should be emphasised that there are three stages in the analysis of bituminous materials and that careful attention must be paid to each stage if a meaningful result is to be obtained. The stages are:

(a) Sampling of quantity of material from a much larger bulk.
(b) Sampling reduction to produce a sample of size required for test.
(c) The actual analysis.

The procedures (a) and (b) are also applicable to the binder tests already described as well as to tests on small samples of other materials such as concrete obtained from a much larger bulk. Careful attention to sampling with bituminous mixtures is particularly important, however, since representative samples are often quite difficult to obtain.

*Sampling*
With the exception of mastic asphalts, there are four points at which samples may be taken. (It is an advantage if sampling tools are lightly oiled.)

1. From a lorry load. Take three increments of material of nominal size 20 mm or smaller or four increments of larger sizes from about 100 mm below the surface of the material. The increments should be of about 7 kg each and as widely spaced as practicable, though not from the sides of the vehicle. Avoid lumps of coarse surface material falling into the holes during sampling.
2. From a mixer. The sample pan should preferably be fixed mechanically to the mixer so that it can be passed accurately through the centre of the stream on discharge. Take three or four increments according to aggregate size, avoiding the beginning and end of the discharge.
3. From the augers (screws) of the paver. Take two increments alternately from each side of the paver while augers are fully charged and in motion.
4. From the laid-but-not-rolled material. (Not recommended for wearing coarse material or where the nominal aggregate size is within 20 mm of the thickness being laid.) Place two 375 mm square steel trays not more than 10 mm deep each side of the centre line of the paver path. The corner of each tray is connected by steel wire to the area clear of the paver. After the passage of the paver, lift the wire to locate the trays and carefully remove. Combine the two samples. (This method has the disadvantage of causing additional inconvenience during the paving operation.)

Mastic asphalts may be sampled from at least six blocks of the material, or during discharge from the mixer, the principles of collection from a mixer being as given above for other materials.

The minimum sample masses required for the various materials are given in Table 7.2. Too large a sample is undesirable, since extra subdivision is then required.

Table 7.2   Minimum size of bulk sample for
different nominal size materials

| Material | Minimum mass (kg) |
|---|---|
| Nominal size larger than 20 mm | 24 |
| Nominal size 20 mm and smaller | 16 |
| Mastic asphalt | 6 |

*Sample reduction*

Riffling and quartering of the received bulk samples may be used for reducing the size to that required for testing, though a riffle box is to be preferred, since errors resulting from quartering are larger. The riffle box may be lightly oiled or heated if sticky materials are being subdivided.

The masses required for analysis are given in Table 7.3.

## Experiment 7.8   Hot extractor method of analysis of bituminous mixtures (BS 598, Part 2)

The purpose of this method is to determine the binder content of a bituminous mixture and to obtain the grading of the mineral fraction. It is especially suitable for the analysis of older materials which are likely to have a significant water content. (The method is not suitable for the analysis of mastic since the filler would block the filter paper. It may take up to two days to carry out correctly, although, provided suitable apparatus is available, it does not require great skill.)

Table 7.3   Masses of bituminous mixtures
required for analysis

| Nominal aggregate size (mm) | Mass of sample (kg) |
|---|---|
| 50 | 3–5 |
| 40 | 2.5–4 |
| 28 | 2–3 |
| 20 | 1–1.5 |
| 14 | 0.8–1.2 |
| 10 | 0.5–1.0 |
| 6 | 0.2–0.5 |

## Apparatus

The sample container is in the form of a cylindrical vessel made from brass gauze of mesh 1–2 mm or equivalent, mounted in the upper part of a flanged pot fitted with an upper cover and gasket. This is connected to a reflux condenser and a 12.5 ml Dean and Stark graduated receiver. An electric hot plate should be used for heating; No. 1 Whatman filter paper; oven, desiccator, weighing balance to 0.05 g accuracy; trichloroethylene; pipette.

## Procedure

Fit a No. 1 Whatman filter paper into the gauze container to form a complete lining; dry at 100–120 °C, cool in a desiccator and weigh ($M_1$). Place the gauze container in the pot. Place a sample of the correct mass (Table 7.3) in the container and weigh to 0.05 g accuracy ($M_2$). Pour adequate trichloroethylene over the sample but do not submerge. Bolt the cover with its gasket in place, connect the receiver and condenser and heat so that two to five drops of condensate fall per second from the end of the condenser onto the sample and through the filter paper. Water will collect in the receiving tube. (If this becomes full, a measured quantity must be removed by means of a pipette, after temporarily stopping the distillation. Alternatively, if a stop cock is provided on the receiver, the water may be drawn off.) Continue heating until extraction is complete and no further water is collected. Measure the total water collected in g ($W$). Remove the aggregate in the container and dry to constant mass at 100–120 °C. Allow to cool and weigh the cylinder and contents ($M_3$). The binder content, expressed as a percentage, is then

$$\frac{M_2 - (M_3 + W)}{M_2 - (M_1 + W)} \times 100$$

Any fine matter which may have have passed through the No. 1 Whatman filter is ignored in the above determination, as it is in the following simplified method for determining the aggregate grading.

Transfer the aggregate from the filter paper to a metal bottle of suitable size, brushing the filter paper gently to remove as much aggregate as possible. Add solvent to the aggregate and wash through a 75 μm sieve under a 1.18 mm protective sieve. Filter the material passing through both sieves using a weighed No. 5 Whatman filter paper. Dry this filter paper and weigh again to give the approximate amount of filler. Combine and dry the remaining aggregate and grade by the same principles that concreting aggregates are graded.

The sieves that are to be used will depend on the type and size of material being tested. (See BS 594 for rolled asphalts specifications and BS 4987 for coated macadams specifications.)

Compare the aggregate grading with those of Fig. 7.2.

## Experiment 7.9    Sieving extractor method of analysis of bituminous materials (BS 598, Part 2)

This method is more rapid than the hot extractor method – it may be completed in 1 to 3 hours. It has, however, two disadvantages:

1.  Shaking material through sieves may break down softer aggregates such as limestone, leading to a misleading grading.
2.  Wet sieving tends to lead to a finer particle grading than dry sieving, which forms the basis of standard specifications.

This test is unsuitable for mastic asphalts.

If water is present, the percentage by mass will be required, using a technique such as that of Experiment 7.8.

### Apparatus
Sieving extractor apparatus comprising a nest of specially constructed sieves, with gaskets, attached to a base by clamps. The apparatus is capable of being rocked by electric motor and has a drain cock for the solvent. A domed top can be fitted above the nest of sieves and clamps are provided so that the whole apparatus is leak-proof. Centrifuge; metal bottle of 2.5 litre capacity with rubber stoppers; recovery apparatus comprising water bath and flat-bottomed flask which can be connected to a vacuum pump fitted with a pressure gauge. Volumetric flasks; dichloromethane; dried silica gel; weighing balance.

### Procedure
Weigh the sample to 0.05 per cent accuracy ($M$). See Tables 7.3 and 7.4 for the correct sample size, sieve sizes to be used and volume of solvent. Note that the deeper sieves (25 mm depth) are required for some materials. Assemble the clean sieves on the base of the sieving extractor with size decreasing downwards and domed head uppermost. Tighten firmly. Add the required measured volume ($V$) of dichloromethane to the sieves. After checking for freedom from leaks, add the sample to the top sieve, together with a weighed quantity of silica gel (say 25 g) to absorb any water which may be present. Shake for 10 min for coarser materials and 20 min for finer materials.

Remove the liquid proof extractor head and run off the solution into a metal bottle. Shaking may continue during this process, provided escape of solution is prevented.

Separate insoluble matter from the binder by centrifuging. Transfer a volume $v$ to a flat-bottomed flask weighed to 0.01 g accuracy.

Place the flask in the recovery apparatus and boil at reduced pressure (not less than 600 mb), shaking so that the binder is deposited as a thin layer on the walls of the flask.

For highly viscous residues, reduce the pressure to 200 mb for 3.5 min. For low-viscosity residues, increase pressure to atmospheric and then reduce to 600 mb for 3.5 min.

Table 7.4 Recommendations for the size of sample, sieves and volume of solvent to be used for test by the sieving extractor method

| Size and type of material | Sample mass (g) | Recommended sieves (12.5 mm and 25 mm deep) | | | | | | | | Minimum volume of solvent (ml) |
|---|---|---|---|---|---|---|---|---|---|---|
| | | (mm) | | | (µm) | | | | | |
| | | 3.35 | 2.36 | 1.18 | 600 | 300 | 212 | 150 | 75 | |
| **40 mm nominal size** | | | | | | | | | | |
| Macadams | 2500–3750 | 3.35 | – | 1.18 | 600 | 300 | – | – | 75 | 3500 |
| Rolled asphalt | 2500–3750 | – | 2.36 | – | 600 | 300 | 212 | 150 | 75 | 3500 |
| **28 mm nominal size** | | | | | | | | | | |
| Macadams | 1500–2250 | 3.25 | – | 1.18 | 600 | 300 | – | – | 75 | 2500 |
| Rolled asphalt | 1500–2250 | – | 2.36 | – | 600 | 300 | 212 | 150 | 75 | 3500 |
| **20 mm and 14 mm nominal size** | 20 mm / 14mm | | | | | | | | | |
| Macadams | 1000–1500 / 800–1200 | 3.35 | – | 1.18 | – | 300 | – | – | 75 | 2000 |
| Dense tar surfacing | 1000–1500 / 800–1200 | 3.35 | – | 1.18 | – | 300 | – | – | 75 | 3000 |
| Rolled asphalt | 1000–1500 / 800–1200 | – | 2.36 | – | 600 | – | 212 | 150 | 75 | 2500 |
| Coated chippings | 2000–3000 / 2000–3000 | 3.35 | – | 1.18 | – | 300 | – | – | 75 | 2000 |
| **10 mm nominal size and smaller** | | | | | | | | | | |
| Macadams | 700–1000 | 3.35 | – | 1.18 | – | 300 | – | – | 75 | 2000 |
| Cold asphalt, coarse | 600–900 | – | 2.36 | – | 600 | – | – | 150 | 75 | 2000 |
| Cold asphalt, fine | 600–900 | – | 2.36 | – | 600 | – | 212 | 150 | 75 | 2500 |
| Mastic asphalt percentage retained 2.36 mm > 5 | 500–750 | – | 2.36 | – | 600 | – | 212 | 150 | 75 | 2500 |
| Mastic asphalt percentage retained 2.36 mm < 5 | 300–500 | – | 2.36 | – | 600 | – | 212 | 150 | 75 | 2500 |
| Rolled asphalt percentage retained 2.36 mm > 10 | 700–1100 | – | 2.36 | – | 600 | – | 212 | – | 75 | 2500 |
| Rolled asphalt percentage retained 2.36 mm < 10 | 700–1100 | – | 2.36 | – | 600 | – | 212 | – | 75 | 2500 |
| Coated chippings | 2000–3000 | 3.35 | – | 1.18 | – | 300 | – | – | 75 | 2500 |

Cool the flask and weigh to 0.01 g accuracy. For an accurate result, 0.75–1.25 g of binder should be left. The soluble binder, expressed as a percentage by mass ($S$), is given by

$$S = \frac{10\,000\,Z\,V}{vM(100-P)} \left[1+\frac{Z}{dv}\right]$$

where

| | | |
|---|---|---|
| $M$ | = | mass of undried sample (g) |
| $Z$ | = | mass of binder recovered |
| $V$ | = | total volume of solvent |
| $v$ | = | volume of solution used for binder recovery |
| $d$ | = | relative density (1.0 for bitumen; 1.15 for tar mixtures) |
| $P$ | = | percentage by mass of water in the sample. |

The remaining aggregate should be washed, using about 750 ml of dichloromethane, the sieves then being dried in a drying cupboard. Material passing the 75 µm sieve (filler) may be measured by filtration while remaining aggregate is graded in the normal way, using sieves appropriate to the material. (See BS 594 for rolled asphalts specifications and BS 4987 for coated macadams specifications.)

## Experiment 7.10  Surface dressing and hot-rolled asphalt; rate of spread of coating chippings (BS 598, Part 3)

This test is meaningless unless carried out from a full-size spreading machine in the course of surface dressing. This is often quite easily arranged with a local authority.

*Apparatus*
Five to ten 300 mm square trays with four attached lifting hooks; suspension chains and spring balance (which may be calibrated direct in kg/m$^2$).

*Procedure*
Number and weigh the trays, using the chains and spring balance and position 5 to 10 trays in the path of the spreader vehicle.

From an experimental point of view, the most convenient method is to run the chippings spreader 'dry' for a short length – that is, without the surface being first sprayed with binder. If this is not possible, trays can be inserted between sprayer and chippings spreader, though they must be debonded from the binder by sheets of paper under each tray.

After passage of the chippings spreader, retrieve the trays and weigh. Determine the average weight of spread of the chippings:

$$\text{Rate of spread} = \frac{\text{Average mass of chippings per tray}}{0.3 \times 0.3} \text{ kg/m}^2$$

The rate should then be converted into m$^2$ coverage per tonne.

Typical rates of spread are:

| | | | |
|---|---|---|---|
| 20 mm chippings | 50–65 | m²/tonne | |
| 14   "   " | 65–85 | " | " |
| 10   "   " | 85–110 | " | " |
| 6.3   "   " | 115–150 | " | " |

## Experiment 7.11    Determination of rate of spread of binder in surface dressing (RN 39, Appendix 4)

This test can only be realistically carried out using a full-scale sprayer in the course of surface dressing operations. This should, however, be fairly easy to arrange with a local authority.

*Apparatus*
Five or more light metal trays 200 mm square by 5 mm deep; weighing balance accurate to 0.1 g; two-pronged fork and stick for tray removal.

*Procedure*
Number and weigh the trays. Place at least five trays on flat horizontal surfaces in the path of the binder distributor, well clear of vehicle wheels or other obstructions.

After the sprayer has passed, remove the trays with the fork, holding each tray onto the fork with the stick. Wrap in weighed sheets of paper and weigh to 0.1 g accuracy. The tray weights will give a good indication of the average rate of spread and of the variation from place to place. The mean rate of three or more trays should not vary from the specified value by more than 10 per cent.

$$\text{The rate of spread} \; = \; \frac{0.025 \times \text{Mass of binder (g)}}{\text{Relative density}} \; \text{litre/m}^2$$

Assumed relative density is 1.0 for cutback bitumen and is 1.15 for tar. Rates of spread depend on the type of surface and traffic load values, varying between 1.4 litre/m² for lightly trafficked, hard surfaces and 1.1 litre/m² for heavy traffic.

## Experiment 7.12    Measurement of the strength performance of asphalts using the Marshall test (BS 598, Part 3)

The Marshall test is an arbitrary test for measuring the effect of binder content on the performance of asphalts for a given aggregate type. In the test, loadbearing capacity and deformation of a partly confined sample are measured (Fig. 7.12). The stresses produced are complex and not representative of the triaxial stresses encountered in practice. Nevertheless, the test is still used and forms part of BS 598. In a full analysis, 12 or more

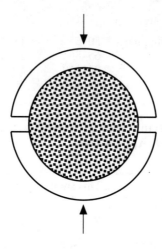

Fig. 7.12   Principle of the Marshall test.

specimens must be made with varying binder contents. Time will normally be against this and the simplest approach is probably to illustrate the principle of the test by the use of one previously manufactured sample.

*Apparatus*
Oven capable of maintaining a temperature in the range 80–200 °C; mechanical mixer – 5 litre capacity, fitted with paddles; the bowl of the heater can be heated by an external purpose-designed electric element; compaction mould required for making specimens; compaction pedestal and hammer; extractor for removing specimens from mould; water bath – thermostatically controlled; balance to weigh at least 2 kg to 0.1 g accuracy and adapted for weighing specimens in water; Marshall load frame.

*Procedure*
The sample should be of correct size to produce one compacted specimen 101.6 mm in diameter by 63.5 mm high.

Heat the sample in the oven at a temperature of approximately 165 °C, depending on binder type, and mix thoroughly in the heated mixer. Place the material in the lightly oiled heated circular mould complete with collar, using tough absorbent paper at the top and bottom to prevent sticking. Apply 50 blows of the compaction hammer and then reverse the mould and collar and apply 50 blows to the other face.

Cool the specimen, in its mould, in water and remove using the extractor. Measure the height and if it is not within the range 63.5 ± 1.5 mm, reject and repeat. Dry and then weigh in air ($W_1$) and again in water ($W_2$) by means of an adapted weighing balance or by suspending from a spring balance. The volume of the specimen is ($W_1 - W_2$) g.

Place the specimen in a water bath at 60 °C for 1 h and then test sideways in compression, using the specially designed testing heads and loading at 50 mm/min. Read the maximum load L.

Obtain the stability correction factor from Table 7.5. Calculate the stability (S):

$$S = L \times \text{Stability correction factor}$$

Calculate the compacted density of mix (CDM):

$$\text{CDM} = \frac{W_1}{W_1 - W_2} \text{ g/ml}$$

If the binder content (A) is known, calculate the compacted density of aggregate (CDA):

$$\text{CDA} = \text{CDM} \left( \frac{100 - A}{100} \right) \text{ g/ml}$$

When designing mixtures, the results of CDM, CDA and S are plotted for various binder contents, those contents giving maximum values of each of the three parameters being determined. The average of these three binder contents is then taken as optimum. Typical values of CDM, CDA and S at optimum binder content are 2.26 g/litre, 2.10 g/litre and 6.0 kN, respectively.

Optimum binder contents are usually in the region of 7 per cent by weight.

Table 7.5  Marshall test: correction factors for stability values with variations in height or volume

| Height of specimen (mm) | Volume of specimen (ml) | Stability correction factor |
|---|---|---|
| 62 | 502–503 | 1.04 |
|  | 504–506 | 1.03 |
|  | 507–509 | 1.02 |
|  | 510–512 | 1.01 |
| 63.5 | 513–517 | 1.00 |
|  | 518–520 | 0.99 |
|  | 521–523 | 0.98 |
|  | 524–526 | 0.97 |
| 65 | 527–528 | 0.96 |

234

## Questions

1. Distinguish between bitumen and tar
   (a) in terms of manufacture
   (b) in terms of properties/uses
2. Comment on:
   (a) the weathering properties
   (b) the temperature susceptibility
   (c) the action of organic solvents on bitumen compared with tar
3. Distinguish between blown bitumen and cutback bitumen. Give basic properties and applications of each.
4. Explain what is meant by bitumen emulsions and describe briefly how they 'break' or set. Hence indicate why ponding should be avoided. Give two applications of bitumen emulsions in construction.
5. Distinguish between asphalts and bitumen macadams as used in road construction. Give the special properties and applications of mastic asphalts.
6. Suggest possible causes of cracking in asphalt roofs and indicate how, by good design and construction, the risk of cracking can be minimised.
7. Distinguish between rolled asphalt and dense bitumen macadam in respect of composition and properties.
8. Explain what is meant by blistering in asphalt or built-up felt roofs. Suggest:
   (a) how the risk of blistering can be reduced
   (b) what remedial steps may be taken
9. Explain the meaning of the term *embedment* in relation to surface dressing on roads. Indicate what circumstances would lead to 'fatting-up' and 'scabbing' in surface dressings.
10. Describe tests which could be used for measuring
    (a) the hardness
    (b) the behaviour at elevated temperatures
    of a bitumen. Explain the relationship between these properties.

## References

Road Note 39, *Recommendations for surface dressing*. Department of Transport, Transport and Road Research Laboratory, 1981.
*The Shell Bitumen Handbook 1990*, Shell UK Ltd.

## British Standards

*Binder*

BS 76: 1974, *Specification for tars for road purposes.*
BS 3690, Part 1: 1989, *Specification for bitumens for roads and other paved areas.*

*Mixtures*

BS 63, Part 1: 1971, *Single-sized roadstone and chippings.*
BS 594, *Hot rolled asphalt for roads and other paved areas.*
Part 1: 1992, *Specification for constituent materials and asphalt mixtures.*
Part 2: 1992, *Specification for transport, laying and compaction of rolled asphalt.*
BS 1446: 1973 (1990), *Specification for mastic asphalt (natural rock asphalt fine aggregate) for roads and footways.*
BS 1447: 1888, *Specification for mastic asphalt (limestone fine aggregate) for roads, footways and pavings in building.*
BS 4987, *Coated macadams for roads and other paved areas.*
Part 1: 1988, *Specification for constituent materials and for mixtures.*
Part 2: 1988, *Specification for transport, laying and compaction.*

*Tests*

BS 598, *Sampling and examination of bituminous mixtures for roads and other paved areas.*
Part 2: 1974, *Methods of analytical testing*
Part 3: 1985, *Methods for design and physical testing.*
Part 100: 1987, *Methods of sampling for analysis*
BS 2000, *Methods of test for petroleum and its products*
Part 47: 1983, *Solubility of bituminous binders.*
Part 49: 1993, *Determination of needle penetration of bituminous material.*
Part 58: 1988, *Determination of the softening point of bitumen. Ring and ball method.*
Part 72: 1988, *Viscosity of cut-back bitumen.*
BS 4689: 1980 (1990), *Method of determination of flash and fire points of petroleum products. Cleveland open cup method.*

**ter 8**

# Plastics

The term 'plastic' refers to the fact that many plastics become soft and malleable when heated, though a more accurate description would be 'polymeric materials' which are almost always of organic origin; that is, based on carbon. These materials have in general the following attractions:

- They have low density – usually around that of water.
- An almost infinite range of plastics can be produced – based on the versatility of the carbon atom in bonding terms.
- Many plastics are quite inert – they do not absorb water and are unaffected by frost, pollutants, dilute acids and alkalis. Hence surface protection from moisture is unnecessary.
- They are readily moulded/shaped.
- They are good thermal insulators, especially in foam form.

Some shortcomings of organic polymers are as follows:

- They are often susceptible to ultraviolet deterioration.
- They have high thermal movement.
- They have low stiffness (elastic modulus).
- They have generally low resistance to heat and some are susceptible to fire, especially in foam or sheet form.
- They rely (currently) on fossil fuels as the raw material.

The technology of plastics is being continually developed and solution of some of the above problems, at least in part, is likely in the foreseeable future.

An understanding of plastics will depend very much upon a study of their chemical make-up and so a short section is devoted first to this.

## The periodic table and the importance of carbon

The very important role of carbon in producing polymeric materials can be seen by reference to the periodic table (Table 8.1). In this table the elements are grouped according to the number of electrons in their outer shells. Except for the lightest elements (hydrogen and helium), elements containing 8

Table 8.1   Simplified periodic table

| Shell no. | Groups | | | | | | | |
|---|---|---|---|---|---|---|---|---|
| | 1 | 2 | 3 | 4 | 5 | 6 | 7 | 8 |
| 1 | 1 H | | | | | | | 2 He |
| 2 | 3 Li | 4 Be | 5 B | 6 C | 7 N | 8 O | 9 F | 10 Ne |
| 3 | 11 Na | 12 Mg | 13 Al | 14 Si | 15 P | 16 S | 17 Cl | 18 A |
| 4 | 19 K | 20 21–30[a] Ca | 31 Ga | 32 Ge | 33 As | 34 Se | 35 Br | 36 Kr |
| 5 | 37 Rb | 38 39–48[b] Sr | 49 In | 50 Sn | 51 Sb | 52 Te | 53 I | 54 Xe |
| 6 | 55 Cs | 56 57–80[c] Ba | 81 Tl | 82 Pb | 83 Bi | 84 Po | 85 At | 86 Rn |

*Notes:*

Shell capacities are:   Shell   1:   2 electrons
Shell   2:   2 + 6 electrons
Shell   3:   2 + 6 + 10 electrons
Shell   4:   2 + 6 + 10 + 14 electrons, etc.

[a]   Elements 21–30. These are the 10 transition elements; comprising scandium, titanium, vanadium, chromium, manganese, iron, cobalt, nickel, copper, zinc. Their properties are rather similar because they correspond to electrons filling orbits in shell 3, shell 4 at this stage having two electrons in it, which largely dictate their character.

[b]   Elements 39–48. These represent the 10 electron orbits in shell 4 filling  in a similar way to the 10 elements as in note *a*, though the elements  are less abundant.

[c]   Elements   57–80. These represent a further 14  electrons filling orbits in shell 4 plus 10 electrons filling shell 5.

electrons in their outer shell have maximum stability and therefore minimum tendency to bond (elements bond with each other in order to minimise their energy). Incomplete shells correspond to high-energy states and result in strong bonding tendency while complete electron shells, corresponding to low-energy states, result in small bonding tendency. Hence all group 8 elements, which have complete shells, have a very low bonding tendency. This results in their having a gaseous form (inert gases), since solids and liquids are produced as a result of bond formation. Elements in all other groups have a tendency to bond, elements in groups 1, 2 and 3 tending to lose electrons and elements in groups 5, 6 and 7 tending to gain electrons. The number of electrons an atom is short, or in excess, of a full shell represents the number of 'bonding opportunities' the atom possesses; for example, sodium (no. 11); potassium (no 19); chlorine (no. 17) and fluorine (no. 9) all have a single bonding opportunity or 'valency'. The transfer of electrons produces a bond, NaCl being a very simple example of the transfer of one electron from the Na to the Cl (Fig. 8.1). Magnesium (no. 12) and oxygen (no. 8) have two bonding opportunities; valency two, and so on.

Carbon is the lightest element with four electrons in its outer shell which gives it great versatility in bonding terms. It usually shares electrons with those of other atoms in order to produce a total of eight electrons. This is easiest with atoms of similar size, for example, H, N, O, F, Cl to give a vast range of compounds normally described as covalent – the result of electron sharing. Silicon (element 14) likewise has four electrons but bonds less readily with lighter atoms because of its larger size. Hence the great majority of plastics are based on C.

To produce a solid material, bonding must be extended between thousands of atoms. Such bonding may result in regular patterns of atoms (crystals); irregular arrangements (amorphous materials) or long chains, which may result in fibres.

## Compounds based on carbon

As indicated above, the carbon atom (no. 6) is four electrons short of a complete shell and generally seeks therefore to share four electrons from

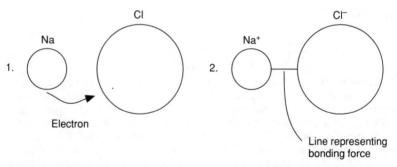

Fig. 8.1   Ionic bonding between sodium and chlorine forming common salt.

other atoms. Bonding with hydrogen to produce hydrocarbons is particularly easy since the hydrogen atoms fit nicely around the carbon atom. The simplest hydrocarbon compound is methane, where one carbon atom is combined with four hydrogen atoms, as shown in Fig. 8.2. Another possibility is for two or more atoms of carbon to combine with hydrogen to form ethane (Fig. 8.3), propane (Fig. 8.4), or butane (Fig. 8.5).

This can obviously be extended indefinitely to form very long-chain molecules, which become heavier as the size increases. It can be seen that the carbon atoms are joined by 'single bonds' corresponding to a single electron shared. Such compounds, where each carbon atom is linked to other atoms by single bonds, are known as saturated compounds. Each of the four carbon bonds is orientated in a particular direction, as shown in Fig. 8.6. The saturated compounds described are known as alkanes or paraffins, the four bonding arms trying to form a pyramid shape with equal angles between them. They have the general formula $C_nH_{2n+2}$. Another way of combining the carbon atoms is in the form of a ring, for example, benzene (Fig. 8.7). Examples of hydrocarbon radicals can be seen in Fig. 8.8 while Fig. 8.9 shows some radicals commonly used to form polymers.

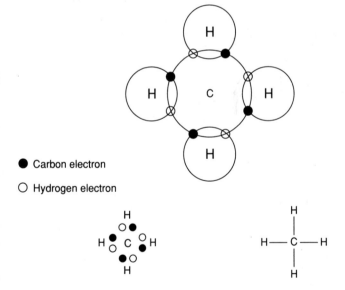

Fig. 8.2 Representations of the bonding between carbon and hydrogen, forming methane, $CH_4$.

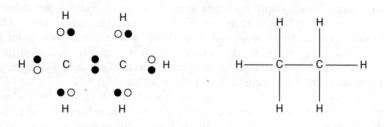

Fig. 8.3    Ethane, $C_2H_6$.

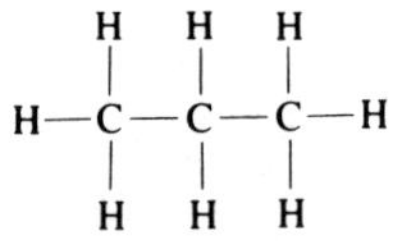

Fig. 8.4    Propane, $C_3H_8$.

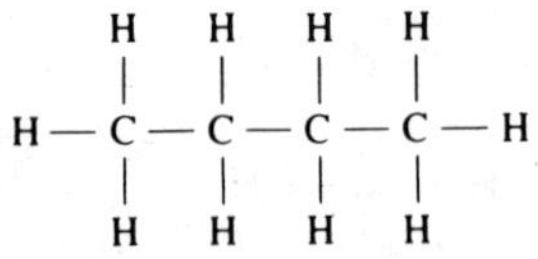

Fig. 8.5    Butane, $C_4H_{10}$.

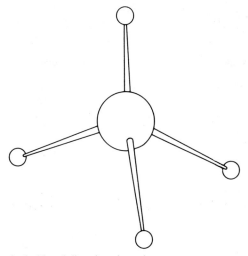

Fig. 8.6    Tetrahedral bond directions in carbon.

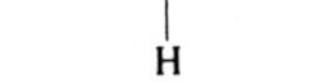

Fig. 8.7    Benzene, $C_6H_6$.

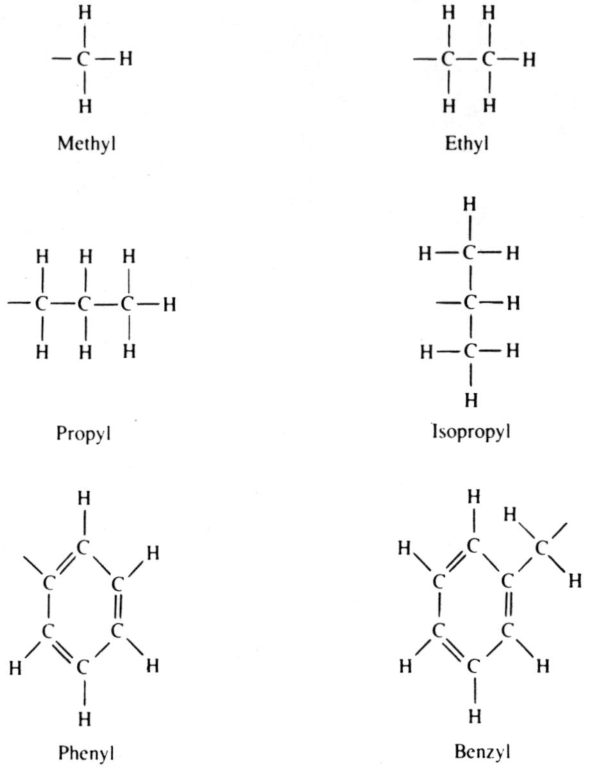

Fig. 8.8   Hydrocarbon radicals.

## Unsaturated compounds

It is possible for two or three electrons to be shared between two atoms, forming double and triple bonds respectively. If an organic compound has one or more of its carbon atoms linked chemically in this way, then it is said to be unsaturated. Ethylene (ethene) is one of the simplest of the unsaturated organic compounds, as shown in Fig. 8.10. Because the four bonding 'arms' of the carbon atom are not spaced in three dimensions evenly around it, this involves considerable stress and is not the most stable arrangement; it would happen, for instance, if there were a shortage of H atoms at the time of chemical combination. An example of a triple bond is shown in Fig. 8.11 – acetylene (ethyne). These multiple bonds in unsaturated compounds are easily broken leaving vacant 'arms' ready to combine with other groups or

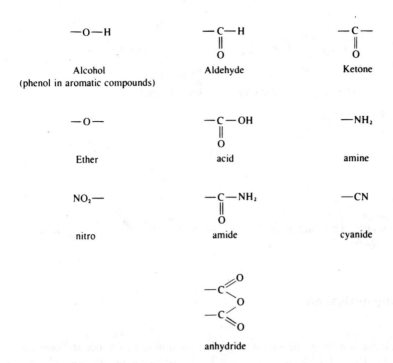

Fig. 8.9   Radicals that combine with the hydrocarbons.

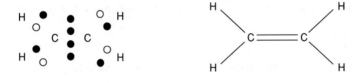

Fig. 8.10   Ethylene (ethene) $C_2H_4$.

Fig. 8.11   Acetylene (ethyne) $C_2H_2$.

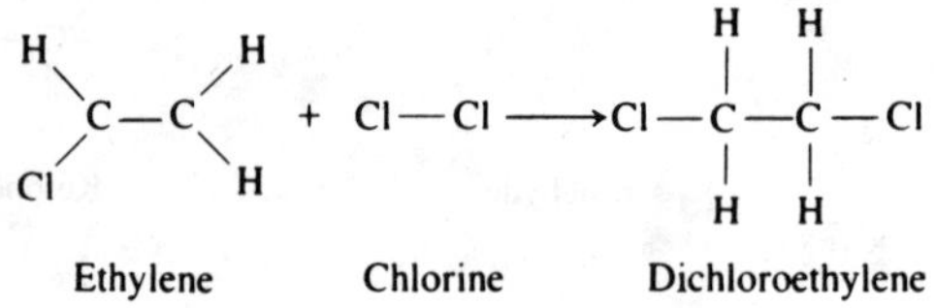

Fig. 8.12 Addition reaction.

elements. This fact is utilised in the production of some very long-chain polymers.

## Polymerisation

Polymerisation is the linking together of simple molecules to form larger ones. The original molecular unit is described as the 'monomer' and the resulting molecule is known as a 'polymer'. The process is accompanied by a gradual change from the liquid state into the solid state.

*Addition polymerisation* (Fig. 8.12) occurs when molecules of a monomer such as ethylene react additively to produce long-chain molecules with possibly as many as 20 000 links in the chain. This can be acheived by means of initiator molecules which effectively contain highly reactive 'free radicals':

$$I\!-\!I \qquad \rightarrow \qquad I^- + I^-$$

Initiator mol.           Free radicals

Introduction of a small amount of initiator into ethylene opens the double bonds:

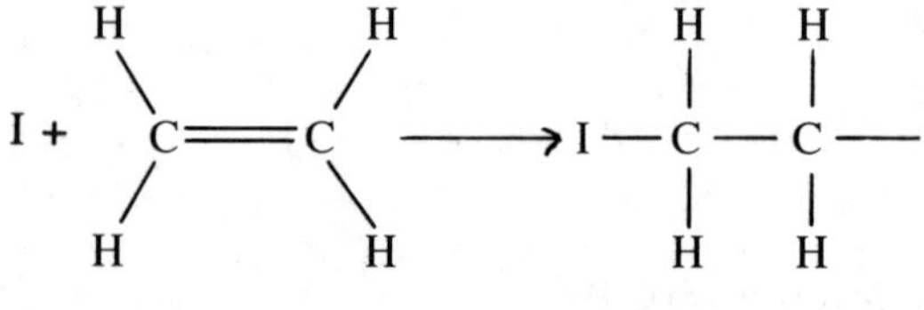

This highly reactive compound can now react with another ethylene molecule:

$$
\begin{array}{ccc}
\underset{\displaystyle H}{\overset{\displaystyle H}{|}}\ \ \underset{\displaystyle H}{\overset{\displaystyle H}{|}} & \ \ \ \underset{}{\overset{\displaystyle H}{\diagdown}}\ \ \ \ \ \underset{}{\overset{\displaystyle H}{\diagup}} & \overset{\displaystyle H\ \ H\ \ H\ \ H}{\underset{\displaystyle H\ \ H\ \ H\ \ H}{}}
\end{array}
$$

I — C — C — + C = C → I — C–C–C–C —

with H atoms above and below each carbon.

The process continues rapidly with molecules continuing to grow until all the ethylene is polymerised, polymer chains terminating in initiator atoms.

*Condensation polymerisation* is a type of polymerisation which usually involves two different types of monomer. The molecules of at least one of these monomers contain two or more reactive groups of atoms or radicals, while the molecules of the other may contain one or more reactive groups. Chemical interaction of these monomers produces a plastics material which has either long-chain molecules or a three-dimensional cross-linked structure, depending on the number of reactive groups, larger numbers being more unlikely to produce three-dimensional molecules. In either case, molecules of such simple substances as water ($H_2O$) or hydrogen chloride (HCl) are eliminated. It is in this respect that condensation polymerisation differs from addition polymerisation. Figure 8.13 shows the polymerisation of phenol and formaldehyde producing 'Bakelite' – one of the earliest plastics to be produced.

## Types of plastics

There are two groups of plastics: thermoplastics and thermosets.

## Thermoplastics

These can be softened by heating after they have been cured, and remoulded, if desired. For manufacturing purposes this has the big advantage that the polymer can be produced in convenient bead form and then supplied to the fabricator to manufacture the product required. In molecular terms, thermoplastics comprise long-chain molecules, lightly bonded by forces arising from charge eccentricities on each polymer chain (van der Waals forces, Fig. 8.14). These forces are overcome by thermal energy on application of heat so that the plastic softens. Problems of bonds of this type are that strength and stiffness are lower than those of more strongly bonded

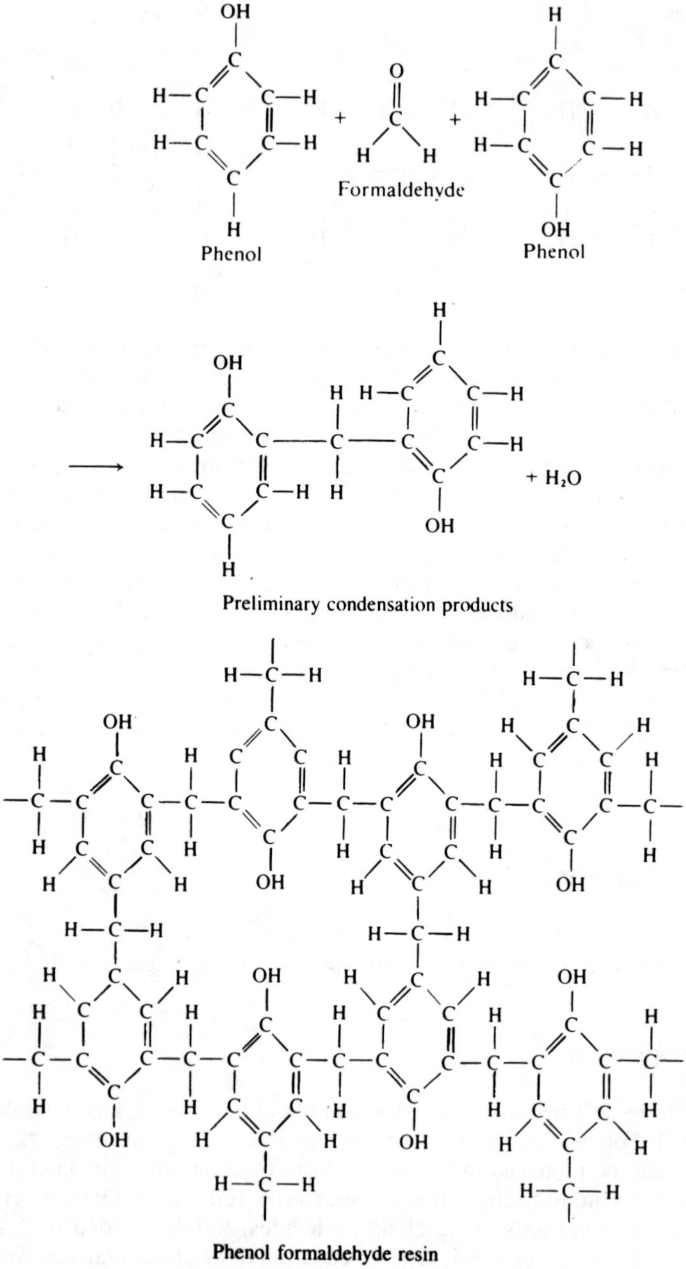

Fig. 8.13   Formation of phenol formaldehyde resin.

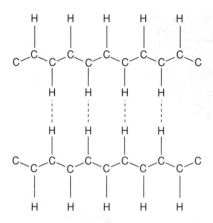

Fig. 8.14   Polyethylene (polythene) chains linked together with Van der Waals bonds (broken lines).

materials, while thermal movement is higher – some polymers have coefficients of thermal expansion over $100 \times 10^{-6}/°C$ (compared, for example, with $11 \times 10^{-6}/°C$ for steel). Nevertheless, thermoplastics are generally cheaper than thermosetting plastics and are more widely used in the construction industry. They can be formed by:

- Injection moulding in which the softened polymer is injected under pressure into a heated mould to form articles of irregular shape such as gutter brackets.
- Extrusion, in which the softened material is forced through a die to give long lengths of constant section, which can be quite complex, producing the maximum possible rigidity for a given polymer type (Fig. 8.15).

Common examples of thermoplastics are polyethylene (polythene), polypropylene, polyvinyl chloride, polymethyl methacrylate (Perspex), polystyrene and nylon. Two important concepts which should be appreciated in relation to thermoplastics are *crystallinity* and *glass transition temperature*.

### Crystallinity
In simple linear polymers, the molecular chains have a tendency to fold back on themselves as they form, chains interleaving and producing relatively dense packing which might be described as a crystal. The result of this is increased bond strength and, hence, increased rigidity for a given polymer. Polymers such as polyethylene, polypropylene, nylon and PTFE crystallise to some extent, their mechanical properties improving as a consequence. Since the crystallites tend to intercept light, these polymers can never be obtained glass-clear; even in thin sheet form they have a milky appearance.

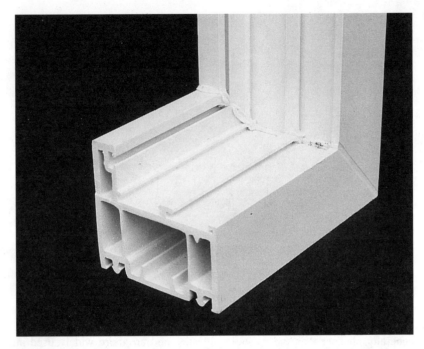

Fig. 8.15 Typical window extrusion in PVC. The joint is made by mitring two sections, heating and then pressing them together. The extrusions are of a complex shape to produce structural stability and to make provision for fixing of rubber sealing strips and steel reinforcement, should it be required.

Polymers with a more random distribution of radicals, or more complex groupings on the carbon backbone, cannot crystallise. These (amorphous) polymers must rely on van der Waals bonds for rigidity, though, in consequence they will be available in glass-clear sheet form (assuming that they do not contain fillers which would interfere with transparency). Most thermoplastics fall into this category.

### Glass transition temperature ($T_g$)

Above a certain temperature, known as the glass transition temperature, the backbone of polymers has sufficient thermal energy to enable the molecules to twist (Fig. 8.16). This involves hydrogen or other side chain groups passing each other, overcoming the repulsive forces which tend to keep them apart. In consequence the flexibility of the polymer increases, and it becomes much more pliable. Conversely, polymers at temperatures below their $T_g$ tend to be stronger and stiffer, but more brittle. Amorphous polymers are generally employed below their $T_g$ value (which is about the same as their softening point), though crystalline polymers, with their extra rigidity due to their crystallites, can be used above their $T_g$.

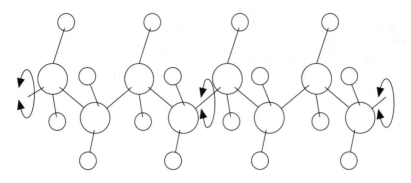

Fig. 8.16   Alternating of bond directions in polyethylene chain so as to minimise repulsion between hydrogen atoms. Above $T_g$ all C—C bonds can rotate as shown.

## Some common thermoplastics

Basic composition, properties and uses of the common thermoplastics are given below. Properties may be compared by reference to Table 8.2 .

*Polyethylene–polythene* $(-CH_2—CH_2-)_n$. This may be regarded as the simplest thermoplastic being a plain hydrogen chain of great length. The structure is highly covalent and in consequence van der Waals forces between the chains are small. The regular chains have, however, a tendency to crystallise, folding back on each other during formation and packing closely together. This produces sufficient cohesion to form a solid, though with a low stiffness and melting point (polythene is above its $T_g$ at room temperature).                                                        Two types are available: high-density polythene which is more crystalline and therefore stronger and stiffer, compared with the low-density material which is less crystalline, though, since both are pure hydrocarbons, each floats on water – that is, they have densities less than $1000\,kg/m^3$. Polythenes have a waxy feel, indicative of the covalent nature of their bonding and cannot be joined by adhesives due to their low surface chemical activity. They are embrittled by ultraviolet light and so must be protected – this is usually achieved by incorporation of carbon black which absorbs the UV rays, avoiding damage. The most important application of low-density polythene is in sheeting for resisting moisture penetration. The material is used both in damp-proof courses and damp-proof membranes. Its toughness in these situations is an advantage, combined with the fact that UV rays should not be a problem. High-density varieties can be used in underground pipes, where their chemical stability and toughness are attractions, though substantial wall thicknesses are essential in view of their low tensile strength.

*Polypropylene* $(-CH_2—CH.CH_3-)_n$. This is similar to polythene except that one H atom is replaced by a methyl group. This leads to higher chain interaction which, together with crystallisation, produces a stiff, stonger polymer with a higher melting point, though, like polythene, it also floats on

Table 8.2 Properties of common thermoplastics, given in order of increasing elastic modulus

| Polymer | Degree of crystallisation (%) | $T_g$ (°C) | Relative density | Expansion coefficient ($\times 10^{-6}$ per °C) | Short-term tensile strength (N/mm²) | Tensile elastic modulus (kN/mm²) |
|---|---|---|---|---|---|---|
| LD Polyethylene | 60 | −120 | 0.92 | 220 | 8 | 0.5 |
| PTFE | 95 | −120 | 2.10 | 110 | 20 | 0.5 |
| HD Polyethylene | 95 | −120 | 0.95 | 130 | 27 | 0.9 |
| Polypropylene | 60 | −27; 10 | 0.90 | 110 | 30 | 1.3 |
| ABS | 0 | 71–112 | 1.05 | 80 | 40 | 2.5 |
| Nylon 66 | 100 | 57 | 1.14 | 90 | 70 | 2.6 |
| Polycarbonate | 0 | 150 | 1.22 | 55 | 60 | 2.7 |
| Polymethylmethacrylate | 0 | 80–100 | 1.18 | 65 | 70 | 2.9 |
| Polystyrene | 0 | 80–100 | 1.05 | 70 | 50 | 3.0 |
| Polyvinylchloride | 0 | 80 | 1.40 | 70 | 50 | 3.2 |

water, being a pure hydrocarbon. It has, like polythene, good chemical resistance and is used for waste systems including laboratory wastes and moulded articles such as manholes. A further important application is as a fibre used to toughen concrete; the drawn fibre has a tensile strength approaching that of mild steel (though the stiffness is much lower) and the fibres appear to act as crack arrestors in brittle materials such as concrete.

*Polystyrene.* This differs from polythene in that it contains benzene rings which greatly increase chain interlock and rigidity, though a slightly brittle material (amorphous, below $T_g$) results. Its most important form is as a rigid or granular insulation material, where its rigidity even in expanded form, enables it to withstand handling stresses. Normal grades are highly flammable and constitute a serious hazard when being stored on construction sites or when fixed in living areas in the form of expanded polystyrene tiles – especially on ceilings. In the latter applications fire-retardant types should be used and tiles should always be fixed with a continuous film of adhesive rather than dabs, which, in the case of fire, would permit the tile to drop easily and 'feed' the fire. A much less hazardous application is in the form of expanded sheeting fixed by adhesive *behind* plasterboard, where the latter affords adequate fire protection.

*Polyvinyl chloride* (PVC) $(-CH_2—CH.Cl-)_n$. Chlorine, with its one electron vacancy, is much more likely to produce ionic bonds, hence an electrically charged chain which in turn produces stronger chain interaction. It also reduces the flammability of the product. Consequently, PVC (amorphous, below $T_g$), with its greater strength and rigidity and reasonable cost, is the most widely used thermoplastic in the construction industry. It has the further advantage of being amenable to solvent welding as well as joining by thermal fusion (Fig. 8.15). It is most commonly used in unplasticised form (uPVC) though organic plasticisers can be introduced to produce softer sheets or fabrics. In common with other thermoplastics it is degraded by ultraviolet light, though stabilisers can be included.

*Polytetrafluoroethylene* (PTFE) $(—C_2F_4—)_n$. Substitution of all the H's in polyethylene for fluorine produces stronger chain attraction which, together with crystallisation, results in a polymer with great chemical stability, low mechanical friction, non-flammability and quite a high melting point.

*Nylons.* These contain amide groups (see page 243) resulting in strong chain attraction, which, together with crystallisation, produces a range of high-performance polymers with melting points over 200 °C.

*Polymethyl methacrylate* (PMMA) (Perspex). This is based on acrylic acid, $CH_2—CH.COOH$, which polymerises to form strongly attracted chains, resulting in a hard, rigid, amorphous polymer. The plastic has excellent optical properties and is used for baths, glazing and shop signs. It is also used in paints, as a binder in resin concrete and for crack repair. Both the resin and the polymer are highly flammable so precautions must be taken where appropriate.

*Acrylonitrile butadiene styrene* (ABS). This is an example of a copolymer combining the rigidity of styrene with the toughness of butadiene styrene rubber. The material is rather similar to PVC, though it is tougher, more expensive and cannot be solvent welded.

Examples of further plastics developed for specific applications include:

*Polyvinyl fluoride* (PVF), for durable plastic coatings.
*Polycarbonates*, tough and rigid glazing.
*Modified PVC products* for hot-water pipes.


## Thermosets

These cannot be softened by heat, hence the products must be moulded prior to polymerisation. This is usually achieved by mixing a resin and hardener and moulding fairly quickly. After hardening, the moulded shape cannot be deformed by heating – the material begins to break down chemically before it softens. Thermosets comprise giant three-dimensional molecules (Fig. 8.13) leading to higher strength and stiffness than thermoplastics, together with better resistance to organic solvents. However, this rigidity often leads to brittleness and unless thermosets are toughened – for example, with fibres – situations in which impact is a possibility should be avoided. Examples are phenol formaldehyde, urea formaldehyde, polyester resins and epoxy resins.

There is in fact no rigid distinction between thermoplastics and thermosets, some polymers being lightly cross-linked to form materials with intermediate properties. Polyesters, for example, can be obtained in a flexible fibre form as used in fabrics or in the much more rigid form as the matrix of GRP (glass-reinforced polyester). Rubbers – another name for elastomers, which comprise zig-zag or helical molecules – can be stiffened by vulcanising, in which chains are cross-linked by means of sulphur atoms.

### Some common thermosets

*Phenol formaldehyde*. Phenolic resins were among the earliest to be produced; they were formerly made from coal tar and given the name 'Bakelite'. After reaction with formaldehyde (HCOOH) the benzene ring in phenol provides three reactive sites at which polymerisation can occur by the condensation reaction shown in Fig. 8.13. A large three-dimensional molecule is produced, leading to a stiff, brittle polymer. All forms are dark brown or black in colour and are used where hardness and heat resistance are required – for example, in electrical goods such as switches and sockets. They are also used in laminates such as 'Formica', paper being used to toughen and reinforce the resin. A further application is in waterproof wood adhesives (for factory use since the hardener is highly acidic).

*Urea formaldehyde*. These resins are similar to phenolic resins except that they are clear so that a large range of colours can be obtained, white electrical goods being the most common application. They are also used, in expanded form, for *in-situ* cavity fills.

*Polyester resins* (GRP). A wide range of thermosetting resins can be made by using styrene to cross-link mixtures of saturated and unsaturated dicarboxylic acids, the precise formulation depending on the degree of rigidity required. Resins are normally partly polymerised to form a syrup and then an inhibitor is added to stop the reaction. At the point of use, polymerisation is then recommenced by addition of a *small* quantity of initiator such as organic peroxide. Since they are brittle on their own they are invariably reinforced with glass fibres which both toughen and stiffen the resin. Glass fibres have an $E$ value of $70 \, kN/mm^2$ compared with about $4 \, kN/mm^2$ for the resin. The performance of the resulting composite, glass-reinforced polyester, or GRP, depends on the proportion of fibres; hand lay-up techniques, as used for car body repairs contain only 10–20 per cent by volume of fibres, but by use of woven roving or winding of unidirectional fibres, $E$ values of up to $40 \, kN/mm^2$ with good tensile properties can be obtained. Polyesters are subject to about 7 per cent shrinkage and this can affect adhesion to other materials. Surface glass fibres should be protected by a 'gel' coat to prevent admission of water which causes deterioration of the fibres. The performance of GRP can only be as good as the protective gel coat which must therefore be properly maintained in an exposed environment. GRP has been used for a wide range of precast moulded products, from inspection chambers to cladding panels and even precast portal frames for swimming pools.

*Epoxy resins.* These contain the epoxide ring:

$$R\!-\!\!\!-\!CH\!-\!\!\!-\!CH_2$$

which is opened to form a polymer in much the same way as a polyester. The final polymerisation process is, however, different from polyesters in that the resin and hardener, having the action of 'hooks and eyes', become linked together in the resultant polymer (Fig. 8.17). It is important to have equal numbers of hooks and eyes, and therefore also important that precisely measured quantities of resin and hardener are mixed. This is best acheived by the use of prebatched containers, carefully labelled – the total contents of the resin and hardener containers being used. Also, unlike polyester resins, most shrinkage is pre-gelation (post-gelation shrinkage only being about 1 per cent), so that good bonding properties are obtained. Epoxy resins have excellent resistance to chemicals, particularly to alkalis, and are used in adhesives ('Araldite'), heavy-duty flooring compositions, resin concretes and glass-reinforced composites. Their high cost is offset by very good performance.

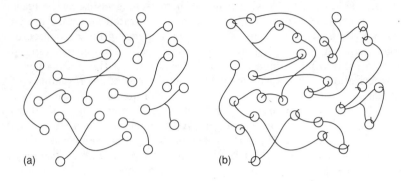

Fig. 8.17  (a) Simple representation of an epoxy resin, epoxide groups being represented by 'eyes'; (b) hardener provides 'hooks' to link up with the 'eyes'. For good results the number of hooks and eyes must be equal.

## Elastomers

The term elastomer is used to describe polymers such as rubbers with a very high strain capability – often well over 100 per cent. They are generally thermoplastics above their $T_g$, which permits bond rotation. Molecules usually form zig-zag configurations so that stretching an arrangement of the type shown in Fig. 8.18 produces increasing resistance initially, followed by much greater resistance as chains approach a straight profile. This behaviour is typical of rubbers – initially they have very low $E$ values, of between 1 and 20 N/mm$^2$ at strains up to 400 per cent. On further stretching, the $E$ value rises as resistance is encountered in the straightened chains, prior to failure. The low $E$ value of elastomers is very important; higher values would generate stress both in the elastomer and in the materials they are used in conjunction with. It is important to appreciate that at very high strain levels materials change shape so that provision must be made for this to occur in use. This, in essence, means avoiding bonding sealants on all sides so that they can be allowed to change shape as movement is encountered (Figs 8.19 and 8.20). Some common elastomers are:

### Natural rubber
This generates very little heat on flexing. It is used for bridge and machine bearings; even entire buildings have been mounted on rubber to isolate the structure from vibrations – for example, from underground trains.

### Silicone rubbers
These are widely used as sealants. They can be obtained as two-part formulations, having the advantage of rapid curing, or as RTV (room temperature vulcanising) types which cure slowly by atmospheric moisture.

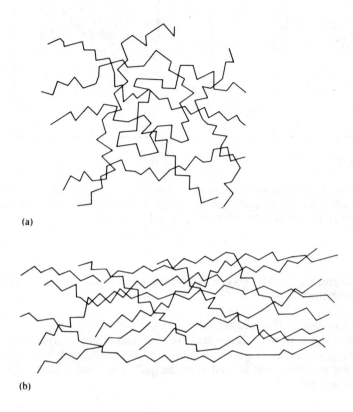

Fig. 8.18 (a) Random zig-zag profile of carbon chain in unstressed elastomer; (b) chain straightening caused by stretching.

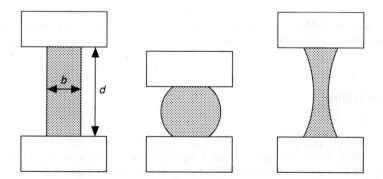

Fig. 8.19 Sealants can only accommodate large movements if they are permitted to change shape. The 'ideal' shape for most purposes is when breadth ($b$) is half depth ($d$).

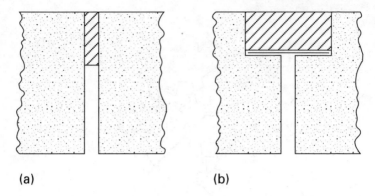

**(a)**                                    **(b)**

**Fig. 8.20**  Incorrect and correct use of elastomers for sealing a crack/joint: (a) will fail because the sealant cannot change shape; (b) the joint is opened out and a debonding strip is used to prevent bonding behind the sealant.

Each is characterised by good adhesion, elastic properties, water repellancy and durability.

*Polychloroprene* (Neoprene)
This has good all-round chemical stability and, since it crystallises on stretching, it combines high extensibility with high strength. It has relatively good fire resistance and is used in contact adhesives and sealing strips for glazing and 'O' rings.

## Application of polymers in the construction industry

Typical applications are given in Table 8.3  together with advantages and precautions applicable to each type.

## Questions

1.  Suggest reasons why plastics have generally low density, referring to Table 8.1 in your answer and suggesting which polymers should have the lowest densities. Check these suggestions against Table 8.2.
2.  Explain why large molecules are most likely to result in a solid state and indicate how the size of simple organic molecules can be increased to produce solids.

Table 8.3  Typical uses of polymers

| Application | Polmer used | Attractions | Possible problem areas |
|---|---|---|---|
| Window frames | Unplasticised PVC | Attractive, maintenance free surface; good thermal insulation; competitive cost; easily made to fit existing opening. | Ensure adequate stiffness, especially if double glazing used; larger sizes should be steel reinforced to avoid creep; colour/finish deteriorate with time |
| Rainwater goods | Unplasticised PVC | Low cost; easy fixing; light weight; range of designs/colours | Low rigidity compared with metal systems; movement joints must be allowed;adequate support needed; possibility of leaks/separation of sections after long service; age may cause embrittlement |
| Waste systems | Unplasticised PVC | Low cost; easily assembled using solvent cement; | Limited resistance to hot effluent/solvents; long lengths may be subject to thermal movement/creep problems; joints cannot be opened after assembly |
| | ABS | Higher temperature/ solvent/impact resistance | Higher cost; special solvent needed for jointing. |
| Glazing | Acrylic plastics | Low cost; high impact resistance; low risk of injury on impact. | Fire risk in some situations; not for use in large glazed areas; scratches easily |
| Cold water cisterns | Glass-reinforced polyester | Light; durable; easily fixed | Good support necessary; avoid stressing at pipe connections. |
| Cavity insulation | Polystyrene bats | Low cost; easily incorporated during construction; cavity can be retained | Some skill/supervision required to ensure correct installation |
| | Polystyrene beads | Can be incorporated into existing cavities | All openings (for example, for services) must be sealed prior to insertion; PVC-coated wiring in cavity may lose plasticiser to beads causing embrittlement; slight risk of damp penetration in severe weather |
| | Urea formaldehyde foam | Easily incorporated at low cost into existing cavities | Not recommended for severe exposure. Foam may crack, leading to moisture admission. Slight risk of formaldehyde gas in interior for a time after insertion |

(continued overleaf)

Table 8.3    *continued*

| Application | Polmer used | Attractions | Possible problem areas |
|---|---|---|---|
| Foam filling for upholstery | Polyurethane foam | Low cost; good comfort level | Serious fire risk; unmodified foams no longer permitted; though much foam still in use. CMHR (combustion modified high resilience) foams now required by law. |
| Fabrics for air-supported /tension structures | Polyester-reinforced PVC | Relatively low cost; material melts in fire venting smoke. | Life limited to 15–20 years |
| | Glass fibre reinforced PTFE | Life up to 50 years; self cleaning | Expensive; material remains intact except in extreme heat- risk of build-up of toxic fumes. |

3. Explain the meaning of the term *glass transition temperature* and its significance in relation to the performance of

   (a) crystalline polymers
   (b) amorphous polymers

   referring to common polymers in your answer.

4. Outline the main attractions and the main disadvantages of plastics in relation to the other major groups of materials, metals and ceramics.

5. Suggest why PVC (polyvinyl chloride) is the most widely used plastic in the construction industry.

6. Discuss the use of unplasticised PVC in window frames, giving recommendations for successful use.

7. Explain how elastomers have such high flexibility, giving recommendations for successful use as joint sealants.

8. Explain why glass fibres are incorporated into glass-reinforced polyester (GRP) and give recommendations for successful use of this composite in cladding panels.

# Chapter 9

# Paints

Paints are surface coatings generally suitable for site use, marketed in liquid form. They may be used for one or more of the following purposes:

- To protect the underlying surface by exclusion of the atmosphere, moisture, fungi and insects.
- To provide a decorative easily maintained surface.
- To provide light- and heat-reflecting properties.
- To give special effects; for example, inhibitive paints for protection of metals; electrically conductive paints as a source of heat; condensation-resisting paints.

Painting constitutes a small fraction of the initial cost of a building and a much higher proportion of the maintenance cost. It is, on this basis, advisable to pay careful attention to the subject at construction stage. Furthermore, there are a number of situations in which restoration is both difficult and expensive once the original surface has failed and weather has affected the substrate; for example, clear film forming coatings on timber, and painting of steel. In other situations, access becomes more difficult later – for example, fascia boards become obscured by gutters. Such situations merit special care.

There are usually three stages in a painting system: primer, undercoat and finishing coat.

## Primer and undercoat

The function of the primer is to grip the substrate, to provide protection against corrosion/dampness and to provide a good key for remaining coats.

The function of the undercoat is to provide good opacity (hiding power) together with a smooth surface which provides a good key for the finishing coat. Undercoats usually contain large quantities of pigment to provide hiding power.

Undercoats and priming coats do not in themselves provide an impermeable dirt-resistant coating.

## Finishing coat

This must provide a durable layer of the required colour and texture. Traditionally most finishing coats were gloss finish and these tend to have the best resistance to dirt since they provide very smooth surfaces. Silk or matt finishes can be obtained if preferred and some paint types such as emulsions will not normally give the high gloss of traditional oil paints.

### Constituents of a paint

The types and proportions of paint constituents tend to be evolved by the manufacturer from experience rather than being designed from 'first principles'. Any one product will be subject to at least small modifications from time to time.

The main components of a paint are the vehicle or binder, the pigment and the extender.

## Vehicle or binder

This is the fluid material in the paint which must harden after application. The hardening process may be due to one of the following:

(a) Polymerisation by chemical reaction with air in the atmosphere. Such paints tend to form a film in a part empty can. They include ordinary 'oil' paints.
(b) Coalescence of an emulsion. Emulsions are pre-polymerised into very small particles which are prevented form coalescing by an emulsifying

agent. They set by water loss leading to 'breaking' of the emulsion.

(c) Evaporation of a solvent. Solvents need to be volatile, hence they are often flammable.

Paints based on types (a) and (b) are described as *convertible coatings* because once set, they cannot easily be re-softened. Once weathered, application of new coatings to these paints therefore relies on the previous paint being roughened to provide a key. Paints based on type (c) are described as non-convertible since they can be re-softened by application of a suitable solvent. Subsequent coats also tend to fuse into previous coats and they do not form films in the can (though the paint may thicken by solvent loss due to evaporation).

The vehicle is largely responsible for the gloss and mechanical properties of the final coating. Vehicles may be blended with:

• driers, which modify hardening properties;
• plasticisers, which increase the flexibility of the hardened film;
• solvents, which adjust the viscosity of the wet paint;
• other additives – for example, with fungicidal action.

## Pigments

These are fine insoluble particles which give the colouring ability and body to the paint. Primers and undercoats tend to have large proportions of pigment to produce opacity, while finish coats have low proportions, since to produce a gloss, the pigment should be beneath the surface. The particle size of pigments is very small in order that maximum colouring power is obtained by minimum thickness of material. Inorganic pigments such as titanium dioxide (white) have the best performance in respect of resistance to solvents, colour fastness and heat resistance, though organic pigments tend to produce the brightest, cleanest colours.

## Extenders

These can be added to control the flow characteristics and gloss of the paint with the added advantage of reducing the cost. Because they are not involved in the colouring process they have a particle size larger than that of the pigment.

## Some common types of paint

### Oil (alkyd resin) paints and varnishes

These are well established and are still the most widely used paints for general purposes including painting of wood and metals. They were traditionally based on linseed oil but modern oil paints are manufactured from alkyd, polyurethane, or other synthetic resins, which allow greater control over flow characteristics, hardening time and hardness/flexibility of the dried film. Once hardened, alkyd resin-based paints behave as thermosetting plastics, being resistant to solution in oils from which they were formed. Unfortunately the hardening process continues slowly with time and these paints tend to become brittle over a period of years, especially if exposed to substantial levels of sunlight. This leads to cracking, especially when they are applied to substrates with high movement tendency such as timber.

### Saponification of oil-based paints and varnishes

Oil- or alkyd-based paints are made by reacting organic acids with alcohols such as glycerol. If an alkali such as calcium hydroxide contacts an oil-based film, there is a tendency to revert to glycerol with the production of the corresponding salt, which in this case is a soapy material – hence the name 'saponification'.

This leads to the breakdown of films and formation of a scum. Hence, oil-containing paints (including polyurethanes) should not be used on alkaline substrates such as asbestos cement, concretes, plasters or renders based on Portland cements, especially when new or if there is a risk of dampness. Alkali-resistant primers, such as PVA emulsion paints, should be applied.

## Emulsion paints

These are now very widely used in interior decorating. Examples are polyvinyl acetate (PVA) emulsion which are suitable for application to new cement or plaster. The molecules are very large but are dispersed in water by colloids to give particles of approximately 1 μm in size. Hence, these paints have the advantage of being water-miscible, although, on drying, coalescence of polymer particles occurs, resulting in a coherent film with moderate resistance to water (Fig 9.1). The film is, however, not continuous, so that the substrate can, if necessary, dry out through the film. Acrylic emulsions for painting of timber have now been produced, including a form which results in a medium gloss finish (see page 265). Emulsions must have a certain amount of thermal energy to coalesce, hence, there is a 'minimum film formation temperature' (MMMF) for each type. Typical MMMF values are, for PVA 7 °C and for acrylic copolymers, 9 °C. They should not be used below these temperatures.

## Cellulose paints

These are solvent-based paints. The cellulose constituent is in the form of nitrocellulose dissolved in a solvent such as acetone. Plasticisers are added to

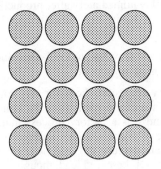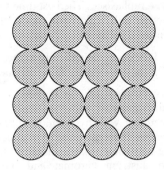

Fig. 9.1 Coalescence of an emulsion to form a coherent but non-continuous film.

give elasticity, and synthetic resins are added to give a gloss, since pure cellulose gives little gloss. Drying usually occurs rapidly, but well-ventilated areas are essential and the paint is highly flammable. Cellulose paints are most suited to spray application (though retarded varieties for brushing are available). These properties, together with the fact that the paints give off a penetrating odour, tend to restrict the use of cellulose paints to factory application. In these conditions, high-quality finishes can be obtained and the resulting coat has good resistance to fungal attack and to chemicals, including alkalis.

## Bituminous paints

These are intended primarily for protection of metals used externally and have poor gloss-retention properties. They are amenable to application in thick coats which therefore give good protection, though the solvents used sometimes cause lifting if applied over oil-based paints, or bleeding in subsequent applied oil-based coats. Sunlight softens the paint, though resistance can be improved by use of aluminium in the final coat. Chlorinated-rubber paints have similar properties. Some uses are based on their resistance to alkalis.

## Painting specific materials

### Ferrous metals

Steel forms the largest bulk of metals used in building and is one of the most difficult to maintain. The need for application to a good substrate cannot be overemphasised. The best time to paint steel is immediately after production, though mill scale (iron oxide film produced during hot rolling) should be removed because:

* It behaves cathodically to the bare metal and may lead to local corrosion.
* It may eventually flake off due to differential movement.

Grit-blasting may be applied to remove any corrosion, though a very rough finish makes it more difficult to achieve a uniform paint film. Pickling – treatment with hydrochloric/phosphoric acid – is a factory process mainly used as a pregalvanising treatment. Inhibitive primers, which interfere with the corrosion process should water be present, include red lead (lead oxide), zinc dust, zinc chromate and zinc phosphate. Of these, red lead, though toxic, is still preferred in 'safe' situations because it is fairly tolerant of poorly prepared surfaces and is amenable to application in thick coats. Metallic lead primers, though non-inhibitive, are fairly tolerant of poor surfaces and may be easier flowing and quicker drying than red lead. They also have superior chemical resistance. On account of their toxicity, lead-based primers are not recommended for use in domestic situations. Where a decorative finish is required, alkyd or aluminium paints may be used. Red oxide (micaceous iron oxide) is a moderately effective primer and is used both in undercoats and finishing coats. Special ('prefabrication') primers about 15–20 μm thick are often applied to steel soon after production to afford weather protection prior to and during fabrication. These have good weathering resistance without the need for further paint coats. They are typically based on phenolic or epoxy resins and may contain etchants and inhibitors.

Generally, the wetter the situation and the more aggressive the climate or atmosphere, the more coats should be given. In extreme situations or where extended life without maintenance is required, protection is only likely to be achieved by impregnated wrappings, bituminous or coal-tar coatings, thick, factory-formed films or prior treatment, such as galvanising.

## Non-ferrous metals

Zinc and aluminium are the non-ferrous metals most likely to require surface coatings and each provides a poor key for paint, unless surface treatment is first carried out. Zinc, in particular, reacts with most oil-based paints, forming soluble salts which reduce adhesion. Zinc should be degreased with white spirit, followed by roughening of unweathered surfaces with emery paper or etching treatment. Primers containing phosphoric acid are available for this; they often also contain an inhibitor, such as zinc chromate. Other suitable primers contain calcium plumbate, zinc dust or zinc oxide. For aluminium, etching is again an advantage, followed by application of zinc chromate or red-oxide primers. Lead-based primers are not suitable.

## Wood

Wood should be protected as soon as possible after the manufacturing process is completed since the surface is quite rapidly affected by weathering/ultraviolet light, as a result of which the paint adhesion properties significantly deteriorate.

Preliminary treatment includes stopping holes and treatment of knots with shellac. A primer is essential to penetrate and yet block the pore structure. Undercoats are unsatisfactory here, since they often do not penetrate the wood and may flake off later. Lead-based primers have been replaced by newer types such as aluminium (BS 4756) and acrylic water-borne primers (BS 5082), as well as the conventional 'solvent based' primers. Acrylic primers are tolerant of higher moisture contents in the timber, though experience shows that permeability is too high for use in single coats as a protection for joinery timber exposed on site prior to installation. Where exposure of primed timber is a possibility, it is recommended that the specification should require that priming paints comply with the appropriate British Standard, since factory-based primers, in particular, often do not comply with the 'six months' exposure test requirements of current standards. Undercoats contribute to the film thickness and therefore protection, though the final coat provides the bulk of the protection. Alkyds and polyurethanes form the basis of most paints, though newer types, such as acrylic emulsions, are now available for external use. Acrylic paints have demonstrated a number of advantages, including:

- rapid drying
- ease of application
- ease of cleaning equipment, brushes, etc.
- good durability, particularly cracking resistance

The cracking resistance is due to the fact that setting does not involve oxygen; hence embrittlement due to continued oxygen penetration does not occur. Possible problems of acrylic paints include the fact that they may inhibit hardening of fresh putty, they may be affected by rain during drying, and gloss levels are not as good as those of conventional alkyd paints.

Some paints are claimed to be microporous or to accommodate wood movement, or both. Tests on microporosity have shown that many such paints do 'breathe', though often not to a markedly greater extent than conventional paints. There is some evidence that, since cracking and blistering are less likely in such paints, they should have greater life. However, it should be appreciated that such paints would not be able to overcome high moisture contents caused by water admission at defective joints or breaks in the paint film. In some cases, there may be a possibility of higher resultant moisture contents in the wood, due to extra penetration through microporous paints during wet spells. Such porous paints produce a 'sheen' rather than the high-gloss finish which may be preferred by some clients. Hence, although there may be advantages in some situations, the need for adequate preparation for painting, together with good maintenance, still applies. The microporosity of paints is also lost if more than about three coats are applied, and this should be considered when redecorating such coatings.

## Varnishes and wood stains

Varnishes are essentially drying oils/resins with little or no pigment, which enhance the natural colouring and grain of timber. Formulations vary greatly, some being suited to external use, though the main problem with all varnishes is that exterior surface maintenance must be meticulous. Cracking or peeling of varnish very quickly leads to bleaching or staining of underlying surfaces due to exposure to water and/or ultraviolet light. Once affected, it is difficult to restore wood to its fomer state. The problem is especially severe on horizontal surfaces and south-facing aspects and it is not, therefore, recommended that varnishes be used in such situations. When varnishes have microporous or other properties such that they are guaranteed not to flake or peel, they can be regarded as stains and may be treated as such. The essential features of stains are:

- They penetrate the wood so that texture and grain remain, though some ('highbuild') types form a surface film in addition.
- They have a low solids content.
- The coatings breathe quite freely, though water repellents are added to reduce admission of liquid water. Fungicides are also usually added to control mould growth.

Moisture contents of stained timber may rise from time to time, such that non-corrosive metal fixings should be used. Since many forms preserve the texture and sometimes the colour of underlying wood, imperfections may be seen and putty is not recommended for glazing – gaskets of glazing beads are preferred. Stains cannot cover cracks or gaps which might be covered by a paint.

An important advantage of wood stains is that coatings generally erode rather than flake or crack, so that maintenance is confined to periodic washing and application of extra coats, though colours become progressively darker with age.

Low-build, transparent forms help retain the natural character of the wood and may leave very little surface film so that protection to wood is correspondingly less and wood may begin to weather comparatively quickly. Many types are now available with coloured pigments which are becoming aesthetically more popular. These increase protection to the wood, though there is the risk that, if a thick surface film is applied, failure characteristics of traditional timber varnishes, cracking and peeling, may occur. Higher-build varieties are recommended on 'difficult' woods, such as some hardwoods, with careful preparation, such as removal of oils, if present. Adhesion is improved on rough-sawn wood finishes, especially of difficult timbers, though these have limited acceptability aesthetically at present.

Not surprisingly, with the variety of finishes obtainable and relative ease of maintenance, the use of wood stains is rapidly increasing.

## Plastics

Most plastics in common use do not require painting, and paint coats, once applied, cannot be removed by normal techniques. Paints, on the other hand, will reduce the rate of degradation of plastics such as polyethylene. Adhesion is poor unless the surface is first roughened to give a mechanical key. The impact strength of some plastics, such as PVC, may be adversely affected, if painted, by migration of solvent into the paint.

## Questions

1. Describe the functions the the primer, undercoat and finishing coat in painting systems. Suggest what might be the effect if

   (a) the primer is omitted
   (b) the undercoat is omitted

   on a wood substrate.

2. Explain the meaning of the term *saponification*, indicating the type of coating in which it is likely to occur, in what situation, and how it can be avoided.
3. Write a specification for a protection system for structural steel in an exposed situation to which people have access.
4. Compare and contrast the following protection systems for wood:

   (a) conventional gloss paints
   (b) microporous paints
   (c) wood stains

## References

### British Standards

BS 4576: 1971 (1991), *Specification for ready-mixed aluminium priming paints for woodwork.*
BS 5082: 1993, *Specification for water-borne priming paints for woodwork.*

# Chapter 10

# Plasters

Plasters may be defined as materials designed to provide a durable, flat, smooth, easily decorated finish to internal walls or ceilings. Plasters were traditionally based on lime or cement but in the last 30 years gypsum has become the most important binder. Gypsum plasters have the following advantages:

- Their setting time can be controlled according to function.
- The time delay between successive coats may be very small.
- Various surface textures and surface hardnesses are obtainable according to function.
- Unlike cement-based plasters, they are non-shrinking (provided plastering technique is correct).
- They have excellent fire-resistance. Set gypsum plaster contains 21 per cent water of crystallisation which absorbs considerable heat, minimising the rate of temperature rise in and behind the plaster. Calcined gypsum also acts as an efficient insulating barrier in fire.

Note that, since gypsum is slightly soluble in water, gypsum plasters are not suitable for exterior uses unless very effective permanent protection is provided.

## The plastering process

A maximum of three coats may be used, each having its own function:

- render coat – to level the background
- floating coat – to produce a flat surface of uniform suction
- finishing coat – to provide a smooth, hard finish

The first two coats require relatively coarse-textured plasters applied in thicknesses of up to 20 mm. The functions of these are now largely combined in a single undercoat or 'browning' coat. The finishing coat utilises a much finer material of thickness up to 5 mm. Some backgrounds, such as plasterboard, provide a flat surface with uniform suction, so that a single finishing coat may suffice.

## Classes of gypsum plaster (BS 1191)

The raw material is calcium sulphate dihydrate ($CaSO_42H_2O$), which is obtained from mines. During manufacture, some or all of the water is driven off by heat, the nature of the resultant material depending on the heating regime. Small amounts of impurities are usually present and these colour the plaster grey or pink, though they have no other significance.

BS 1191 classifies gypsum plasters as follows:

*Class A – hemihydrate* ($CaSO_4^{1}/_2H_2O$: Plaster of Paris)
This is produced by heating to a temperature not in excess of 200 °C. Plaster of Paris sets within 5–10 minutes of adding water, which is far too rapid to permit use in ordinary trowel trades. It is nevertheless useful for moulding purposes such as in decorative plasterwork.

*Class B – retarded hemihydrate* ($CaSO_4^{1}/_2H_2O$)
These are produced from class A plasters by the addition of a suitable set retarder such as keratin. The amount of retarder added depends on the function.

Undercoat plasters tend to be slower setting than finishing coat plasters to allow time for straightening. These plasters are normally designed to be used with sand in ratios of up to 3 : 1 sand : plaster. Increasing the sand content has the effect of accelerating the set, setting times being in the range 2–3 h.

Finishing coat plasters are designed to be used neat or with the addition of up to 25 per cent by weight of hydrated lime, which accelerates the set. Setting times are in the range 1–1$^{1}/_2$ h.

Class B plasters should not be retempered once setting has commenced.

A most important application of class B plasters is in premixed plasters containing lightweight aggregates which are now very widely used. 'Board finish' plasters for plasterboard are also class B.

*Class C – anhydrite*
Class C plasters are obtained by heating the raw material to a higher temperature than class B plasters, with the result that a proportion of

anhydrous calcium sulphate forms, although some hemihydrate also remains. The anhydrous component is so slow to set that an accelerator such as alum is added. The presence of two compounds gives class C plasters a 'double set' – an initial set and then a slow final set, so that the material can be retempered with more water during the first half an hour after adding the water, though it should be applied as soon as possible after mixing and sets completely in $1^1/_2$ h. An important current use of class C plasters is a finishing coat on a sand/cement backing (trade name 'Sirapite').

*Class D – anhydrite* (Keens's cement)
This is harder burnt than class C, resulting in a higher proportion of anhydrite. The plaster set must again be accelerated and stiffening, as in class C plasters, is continuous. The final product has superior strength, smoothness and hardness compared with other types and is used for arrises and surfaces such as squash court walls, where a very durable finish is required. It also provides an ideal base for gloss paints. On account of its strength, it should not be used on soft backings such as plasterboard or fibreboard.

## Lightweight aggregates

Low-density aggregates, such as expanded perlite (produced from siliceous volcanic glass) and exfoliated vermiculite (produced from mica) are a most important ingredient of modern plasters, which are now almost always premixed. Some advantages of these plasters are as follows:

- Transporting and handling costs of the plaster are reduced.
- The low-density fresh material requires less effort to mix and apply and can be used in thicker coats without sagging.
- The thermal insulation of walls or ceilings is improved and the internal surface temperature increased, thereby improving $U$ values and reducing the risk of surface condensation and pattern staining.
- Fire performance of structures is improved.

## Lime

Lime may be used in quantities up to 25 per cent of gypsum in finishing coats of ordinary class B and C plasters. It improves the working properties ('fattiness') of the fresh material and in class C plasters counteracts acidity due to accelerators, hence it may help reduce corrosion of embedded metals. Non-hydraulic limes must be used and should be soaked in water for one day before use.

## Factors affecting the choice of plaster

These may be conveniently described under the headings undercoat and
finishing coat, although the two components are not completely independent
– for example, strong, hard finishing coats also require a reasonably strong
undercoat.

## Undercoat

The most important factor affecting choice of undercoat is background
suction – that is, the tendency of the background to absorb water from the
plaster coating.

Some suction is desirable, since it removes excess water from the plaster
and therefore initiates stiffening. The pores which are responsible for suction
also increase adhesion of the plaster by a mechanical keying effect.

Excess suction is a disadvantage because it results in premature
stiffening of the plaster, giving too little time for levelling and may lead to
poor adhesion.

Very low suction is equally a problem, owing to lack of pores which
improve adhesion.

Table 10.1 indicates plasters which may be used for backgrounds of
varying types. The properties of premixed lightweight undercoat plaster can
be varied according to function – for example, 'bonding' plaster contains
vermiculite aggregate and produces a relatively dense plaster with good
adhesion; 'browning' plaster contains perlite; while 'high-suction
background' plaster contains, in addition, cellulose additives to improve
water retention.

Table 10.1 Undercoat plasters and their applications

| Suction | Examples | Lightweight gypsum plaster |
|---------|----------|---------------------------|
| Low | Dense concrete. No-fines concrete. Engineering bricks | Bonding |
| Medium | Ordinary clay bricks | Browning |
| High | Some stock bricks. Aerated autoclaved concrete blocks | High suction background browning |

Raking of joints in low-suction brickwork or hacking of dense concrete will improve adhesion to backgrounds in the low-suction group. Treatment with PVA-bonding agents both reduces water absorption of high-suction backgrounds and improves adhesion of low-suction backgrounds such as high-strength concrete. Cement/sand/lime plasters form an alternative group of undercoat materials, though these have the disadvantage that the finishing coat application must be delayed for some days to allow curing and shrinkage of the undercoat. One possible advantage of cement-based plasters is that they form a barrier to efflorescent salts if these are present in substantial quantities in the background. It is important, when using cement-based undercoats, that their strength does not exceed that of the background, otherwise shrinkage may result in breakdown of the background surface.

## Finishing coat

This will be selected primarily according to the surface hardness requirement, which will in turn depend on the situation within the building and the function of the building. The most demanding situations are projections in corridors and doorways of public buildings for which the hardest plasters are required. Table 10.2 indicates the gypsum plaster type most appropriate to various situations. Of the plasters listed, the lightweight variety form the largest part of the current market. These do not have the hardness of dense plasters but are nevertheless quite resilient, since impacts are absorbed by localised indentation of the lightweight undercoat.

## One-coat plasters

Plasters are now available which serve as both undercoat and finishing coat. A typical product consists of a white, high-purity Class B gypsum plaster combined with perlite and other additives. The material is best machine

Table 10.2  Finish coat plasters and their applications

| Function | | Plaster type | Comments |
|---|---|---|---|
| Very hard smooth surface | | Class D (anhydrite) finish plaster | Stong undercoat needed. Lightweight types not suitable |
| Hard surface | | Class C (anhydrite) finish plaster | Normally on cement/sand or class B/sand backing |
| Ordinary purposes | Max. fire resistance and insulation | Lightweight class B plaster | Used on lightweight undercoat |

mixed. After application, the material is straightened and then left for about 1 h to stiffen. The surface is then wetted and trowelled with a wooden float to bring fine material ('fat') to the surface. After a further delay, the plaster can then be trowelled to a smooth finish with a metal float. The cost of the material is considerably higher than that of the more common gypsum plasters but a saving in time can be acheived since 'float' and 'set' functions are obtained in a single coat. The background should be of uniform suction (that is, of the same material type) to avoid the difficulty of different areas of the material stiffening at different rates.

## Plasterboards

These consist of an aerated gypsum core sandwiched between and bonded to strong paper liners. Most boards have one ivory-coloured surface for direct decoration and one grey-coloured surface, which has better adhesion properties for plastering. Plasterboards with a foil backing for improved thermal insulation or with a polythene backing for improved vapour resistance are also obtainable.

Plasterboard can be easily fixed to timber studding by nails or special screws, the latter being more satisfactory where the timber frame is of low stiffness. It can also be fixed to masonry, using dabs of special plaster-based adhesives, provided the background is sufficiently straight. In such applications it offers the advantage of requiring only a very small quantity of water; hence it could be used in situations where it is important to keep the relative humidity of the atmosphere to reasonably low levels, where rapid drying is required, or where poor drying conditions prevail. Plasterboard is also available with polystyrene or glass fibre insulation bonded to it, for application either to timber frames or masonry.

A variety of sizes is obtainable (Table 10.3). Laths and baseboards are specifically designed for a plastered finish and, provided joints are staggered, they are less likely to result in cracks at joints than wallboards. Planks are intended primarily for fire-resistance applications. Various edges are obtainable according to function – for example, tapered edges are suitable for smooth seamless joints on boards to be decorated direct (dry lining), square edges are suitable for plastering, and rounded edges, as in laths, give a good bond to the filler which is used between adjacent pieces. For plastered finishes, all joints except those in laths must be reinforced with some form of scrim tape to prevent cracking. Correct procedure is essential in respect of joint treatment to obtain the best finished effect and to avoid cracking.

Where plastering is to be carried out, this may be in one or two coats and a neat class B 'board finish' plaster is normally used. The thickness applied is quite small – in the region of 5 mm – and it is important that drying out is prevented until setting is complete, otherwise a soft, powdery surface will result.

Table 10.3    Examples of thickness and sizes obtainable in gypsum plasterboards

| Type | Length (mm) | Width (mm) | Thickness (mm) |
|------|-------------|------------|----------------|
| Laths | 1200 | 406 | 9.5 or 12.7 |
| Baseboards | 1200 | 914 | 9.5 |
| Wallboard | 2400 | 1200 | 9.5 or 12.7 |
| Plank | 2400 | 600 | 19.0 |

In spite of their paper surfaces, plasterboards provide good fire protection, being designed as class O in Building Regulations.

In terms of impact resistance, plasterboards are not as good as traditional *in-situ* plasters, especially if thin sheets are employed on hollow backgrounds. Thicker sheets or more frequent fixings at least partly overcome such problems. Care is also necessary when making fixings to the material, though reasonably large loads can be carried using special metal or plastic inserts in conjunction with timber battens which spread the load.

## Dry lining

In this technique, the ivory face of plasterboard is used as the finished surface, the only 'wet' trade involved being treatment of joints, nail indentations and any small defects. This offers a further advantage in terms of moisture input into the building. The process cannot, however, be 'rushed'; to obtain a good result, joints must be taped in the same way as when 'skimming' the whole surface and filling must be carried out in several coats, since the filler material can only be applied in small thicknesses. The final finish is also not as hard as that obtained by application of a thin layer of plaster to the entire surface.

## Common defects in plastering

These may be associated with background problems, inadequate or incorrect surface preparation, incorrect use of materials or incorrect plastering technique. The main problems are listed below.

## Cracking

Cracks occur when the plaster is subjected to movement in excess of its strain capacity. Gypsum plasters are non-shrinking but may still be subject to tensile stress if the background moves. Examples are:

1. *Background shrinkage.* This may result if the background is very wet when the plaster is applied. The plaster then cracks at concave corners or following a definite line of cracks in the background. The only practicable remedy is to fill the cracks and disguise with a suitable decorative finish.

2. *Undercoat shrinkage.* Cement-based undercoats may form numerous hairline cracks if not given sufficient time to shrink before the finishing coat is applied. An excess of lime in the finishing coat may have the same effect. These cracks may be filled or simply obscured using a suitable wallpaper.

3. *Plasterboard finishes.* This type of cracking is widespread, usually following plasterboard joints. The extent of cracking is reduced by use of laths or baseboard, movement then being spread over more joints than in larger sheets. Causes include undersized joists, inadequately fixed or restrained joists, poor nailing technique, omission of scrim tape, inadequate joint filling or simply severe impact or vibration of the structure. Fine cracks can be covered by a textured finish but deeper cracks should be cut out and filled. Where background movement is the cause of the cracking, it will be likely to recur.

4. *Structural movement.* This leads to well-defined cracks which follow a continuous line through the structure, plaster cracking in the same position on each side of solid walls. Cracks tend to be concentrated at weak points such as above doorways or windows. Before cutting out and filling, it is essential to establish and rectify the cause of the movement.

## Loss of adhesion

This results when a strong finishing coat is applied to a weak backing coat, especially if the backing is inadequately 'scratched' to form a mechanical key or if it is still 'green'. The problem is uncommon, except with sand/cement backings. Both coats must be replaced unless the problem is caused by a green backing; if this is the case, it should be allowed to harden, loose material removed and then the finishing coat reapplied.

The problem will occur with plasterboard if too much lime is added, or in two-coat work if too much sand is used in the undercoat. The plaster must be stripped off and replaced. If the exposed surface is damaged or uneven, the plasterboard must also be replaced.

## Dry out

This occurs if plaster dries before the water becomes chemically bound by setting. It occurs if thin coats are applied to dry backgrounds such as plasterboard, especially in hot, dry weather. The result is a soft powdery surface which may be difficult to paint or paper. Defective plaster should be stripped and replaced.

## Efflorescence

British Standard 1191 limits the amount of efflorescent salts in gypsum plasters but salts may be present in the background or in sand (when used) if not clean. Salt deposits may appear on drying of plaster, especially if the background is wet during plastering. Deposition may also result subsequently if a plastered area becomes wet due to a leak. The salts crystallise below the surface of emulsion paints, causing loss of adhesion. They can be removed by brushing, once plaster is dry, and will not recur provided the plaster does not become rewetted.

## Questions

1. Give four advantages of gypsum plasters compared with a lime or cement alternative.
2. Describe the composition, properties and uses of the four main classes of gypsum plaster.
3. Suggest why lightweight premixed gypsum plasters have become very popular in recent years.
4. Explain how the choice of plaster for a given situation will depend on:

    (a) the suction of the background
    (b) the requirements of the finished surface

5. Explain the reasons for traditional plastering being carried out in three stages. Give situations in which two- or one-coat work may be acceptable.
6. (a) Describe the various types of plasterboard which are available, giving applications of each.
    (b) Give reasons for cracking in skim coats to plasterboard and suggest how the risk of cracking can be minimised.
7. Give possible causes of adhesion of the finishing coat on:

    (a) solid backgrounds
    (b) plasterboard backgrounds

    Give steps which may be taken to avoid and to rectify the problem in each case.

# References

*British Gypsum White Book: Technical manual of building products.*

## British Standard

BS 1191: 1973, *Specification for gypsum building plasters.*

# Answers to numerical questions

## Chapter 2

**8.** The sand complies with the medium and fine gradings of BS 882 (1992).
**10.** (a)  5.1 per cent
    (b)  9.3 per cent
**11.** Compacting factor incorrect.
**13.**  1(a)  0.47      1(b)  0.58      1(c)  0.52
     2(a)  0.58      2(b)  0.47      2(c)  0.42
     3(a)  0.53      3(b)  0.57      3(c)  0.46
**14.** Water             9 litres
    Cement            19.5 kg
    Sand              29 kg
    Coarse aggregate  61 kg
**15.** Water             1.4 litres
    Cement            3.5 kg
    Sand              4.4 kg
    Coarse aggregate  15.7 kg
**16.** Mean = 25.2 N/mm$^2$
    Standard deviation = 3.6 N/mm$^2$
    Characteristic strength (2 per cent failures) = 17.8 N/mm$^2$
    The required characteristic strength is not being reached

17. The strength requirement of 25 N/mm$^2$ with 5 per cent permissible failures is being satisfied.
18. Water/cement ratio = 0.52

## Chapter 3

4. The brick is of Engineering 'A' classification.

## Chapter 4

3. Porosity = 67 per cent
4. Solid density = 2364 kg/m$^3$

## Chapter 5

6. Carbon equivalent = 0.473 per cent

## Chapter 6

8. (a) Rej    (b) GS    (c) Rej
   (d) SS    (e) GS    (f) GS
   (g) SS    (h) GS    (i) SS
10. (a) Actual stress = 3.67 N/mm$^2$ (satisfactory)
    (b) Deflection = 4.6 (within permissible limits)
11. Applied stress is 1.1 N/mm$^2$. Maximum allowable stress is 1.6 N/mm$^2$. Hence the partition is of sufficient strength.

# Index